Yevgen Gorash

Development of a creep-damage model for a wide stress range

Yevgen Gorash

Development of a creep-damage model for a wide stress range

Application to non-isothermal long-term strength analysis of high-temperature components

Südwestdeutscher Verlag für Hochschulschriften

Impressum / Imprint
Bibliografische Information der Deutschen Nationalbibliothek: Die Deutsche Nationalbibliothek verzeichnet diese Publikation in der Deutschen Nationalbibliografie; detaillierte bibliografische Daten sind im Internet über http://dnb.d-nb.de abrufbar.
Alle in diesem Buch genannten Marken und Produktnamen unterliegen warenzeichen-, marken- oder patentrechtlichem Schutz bzw. sind Warenzeichen oder eingetragene Warenzeichen der jeweiligen Inhaber. Die Wiedergabe von Marken, Produktnamen, Gebrauchsnamen, Handelsnamen, Warenbezeichnungen u.s.w. in diesem Werk berechtigt auch ohne besondere Kennzeichnung nicht zu der Annahme, dass solche Namen im Sinne der Warenzeichen- und Markenschutzgesetzgebung als frei zu betrachten wären und daher von jedermann benutzt werden dürften.

Bibliographic information published by the Deutsche Nationalbibliothek: The Deutsche Nationalbibliothek lists this publication in the Deutsche Nationalbibliografie; detailed bibliographic data are available in the Internet at http://dnb.d-nb.de.
Any brand names and product names mentioned in this book are subject to trademark, brand or patent protection and are trademarks or registered trademarks of their respective holders. The use of brand names, product names, common names, trade names, product descriptions etc. even without a particular marking in this work is in no way to be construed to mean that such names may be regarded as unrestricted in respect of trademark and brand protection legislation and could thus be used by anyone.

Coverbild / Cover image: www.ingimage.com

Verlag / Publisher:
Südwestdeutscher Verlag für Hochschulschriften
ist ein Imprint der / is a trademark of
OmniScriptum GmbH & Co. KG
Heinrich-Böcking-Str. 6-8, 66121 Saarbrücken, Deutschland / Germany
Email: info@svh-verlag.de

Herstellung: siehe letzte Seite /
Printed at: see last page
ISBN: 978-3-8381-3948-7

Zugl. / Approved by: Halle (Saale), MLU, Diss., 2008

Copyright © 2014 OmniScriptum GmbH & Co. KG
Alle Rechte vorbehalten. / All rights reserved. Saarbrücken 2014

PREFACE

The given Ph.D. thesis was accomplished as the final result of my post-graduate study at the chair "Dynamics and Strength of Machines" (National Technical University "KhPI", Kharkiv, Ukraine) and the subsequent academic probation at the chair "Technische Mechanik" (Martin-Luther-Universität Halle-Wittenberg) within the framework of the DAAD scholarship for post-graduate students and young scientists (A/06/09452).

I express my deep gratitude to German Academic Exchange Service (Deutscher Akademischer Austausch Dienst) for the essential financial support during my residence in Germany while working under the Ph.D. thesis and all previous visits within the framework of students academic exchange programs. Thanks for the great opportunity to establish an efficient scientific collaboration with German academics, to realize personal academic development and to achieve the formulated scientific objectives including the defence of Ph.D. thesis.

The given work was created under the scientific supervision and management of Professors Holm Altenbach (MLU Halle-Wittenberg) and Gennadiy I. Lvov (NTU "KhPI", Kharkiv, Ukraine). I would like to thank them for the educating me as a scientist, invaluable academic experience in the solution of complex problems, sufficient freedom of choice and decision, strong motivation and generous support during all the time of the Ph.D. thesis preparation.

The profound gratitude to PD Dr.-Ing. Konstantin Naumenko for the everyday kind assistance in academic problems and effective collaboration in scientific sphere. It would be impossible to complete my Ph.D. thesis without deep theoretical knowledge in the Creep Mechanics and practical skills in graphical design and organization of scientific activities, which Dr. Naumenko generously sheared with me.

Moreover, I appreciate sincerely to all co-workers, Ph.D. students and students at the chair "Technische Mechanik" of Martin-Luther-Universität Halle-Wittenberg for their permanent readiness to help, constructive interest and discussions, favourable team atmosphere and perfect working conditions.

My special heartfelt gratitudes to my parents for their understanding and compliant assistance in all my initiatives and to my beloved fiancée Nadiia Maiboroda for the every day essential encouragement and creative inspiration for the achievement of maximum goals.

Yevgen Gorash   Halle (Saale), July 2008

ABSTRACT

The structural analysis under in-service conditions at various temperatures requires a reliable creep constitutive model which reflects time-dependent creep deformations and processes accompanying creep like hardening and damage in a wide stress range. The objective of this work is to develop a comprehensive non-isothermal creep-damage model based on transitions of creep and long-term strength behavior in a wide stress range. The important features of the proposed creep and damage equations are the response functions of the applied stress which should extrapolate the laboratory creep and rupture data usually obtained in tests under increased stress and temperature to the in-service loading conditions relevant for industrial applications. The study deals with four principal topics including the basic assumptions of creep constitutive modeling, the conventional isotropic and anisotropic creep-damage models, the comprehensive non-isothermal creep-damage models for a wide stress range. Finally, the application to structural analysis of benchmark problems and engineering components is demonstrated.

Within the framework of the constitutive modeling we discuss different creep-deformation mechanisms depending on stress and temperature level and their unification in the frames of one creep constitutive model based on micro-structural experimental studies. An overview of conventional approaches to phenomenological creep modeling with temperature dependence is given. It includes creep-damage models based on the Kachanov-Rabotnov-Hayhurst concept and creep material models with both the initial and the damage induced anisotropy.

The proposed non-isothermal creep-damage model for a wide stress range is based on several assumptions derived from creep experiments and microstructural observations for various advanced heat resistant steels. The constitutive equation affects the stress range dependent behavior demonstrating the power-law to linear creep transition with a decreasing stress. To take into account the primary creep behavior a strain hardening function is introduced. To characterize the creep-rupture behavior the constitutive equation is generalized by introduction of two damage internal state variables and appropriate evolution equations. The description of long-term strength behaviour is based on the assumption of ductile to brittle damage transition with a decrease of stress. Two damage parameters represent the different ductile and brittle damage accumulation. The creep constitutive and damage evolution equations are extended including the temperature dependence using the Arrhenius-type functions. The unified multi-axial form of the creep-damage model for a wide stress range is presented. A new failure criterion includes both the maximum tensile stress and the von Mises effective stress. The measures of influence of the both these stress parameters are dependent on the level of stress.

Several isotropic and anisotropic creep-damage models are applied to the numerical structural analysis using FEM-based CAE-software ABAQUS and ANSYS. These models are incorporated into the software finite element code by means of a user-defined material subroutines. To verify the subroutines several creep benchmark problems are developed and solved by special numerical methods. The examples of long-term strength analysis for various industrial components are highlighted to illustrate the effective features and importance of the continuum damage mechanics approach for the life-time assessments in structural analysis. Finally, an example of long-term strength analysis for the housing of a quick stop valve usually installed on steam turbines is presented. The results show that the developed approach is capable to reproduce basic features of creep and damage processes in engineering structures.

ZUSAMMENFASSUNG

Eine Strukturanalyse unter Betriebsbedingungen bei verschiedenen Temperaturen setzt ein zuverlässiges Kriechkonstitutivmodell, welches zeitabhängige Kriechdeformationen widerspiegelt und begleitende Prozesse wie Verfestigung und Schädigung über einen großen Spannungsbereich erfassen kann, voraus. Das Ziel dieser Arbeit ist die Entwicklung eines umfassenden, nicht-isothermen Kriech- und Schädigungsmodells, das die Übergänge im Verhalten des Kriechens und der Langzeitfestigkeit beschreibt. Wesentlicher Bestandteil der vorgestellten Kriech- und Schädigungsgleichungen sind die Antwortfunktionen zur aufgebrachten Spannung, welche die experimentellen Kriech- und Versagensdaten, die gewöhnlich unter erhöhten Spannungen und Temperaturen ermittelt werden, auf die Betriebsbedingungen extrapolieren sollten. Diese Arbeit ist gegliedert in vier Themenbereiche: Grundannahmen der Modellierung, konventionelle isotrope und anisotrope Kriech- und Schädigungsmodelle, nicht-isothermes Kriech- und Schädigungsmodell für große Spannungsbereiche sowie mehrere Beispiele für die numerische Abschätzung des Kriechens und der Schädigung bei Benchmarkproblemen und Bauteilen.

Im Rahmen der Modellierung werden verschiedene Mechanismen der Kriechdeformation, die vom Spannungs- und Temperaturniveau abhängen sowie deren Beschreibung in einem Kriechkonstitutivmodell, dem experimentelle Ergebnisse zu Grunde liegen, diskutiert. Es wird ein Überblick über konventionelle Ansätze zur phänomenologischen Modellierung des Kriechens mit Temperaturabhängigkeit gegeben. Das beinhaltet Kriech- und Schädigungsmodelle, die auf den Arbeiten von Kachanov-Rabotnov-Hayhurst beruhen, und Kriechmodelle mit sowohl einer anfänglichen als auch einer durch Schädigung induzierten Anisotropie.

Dem vorgestellten Kriech- und Schädigungsmodell liegen Annahmen zu Grunde, die aus experimentellen Beobachtungen für zahlreiche warmfeste Stähle abgeleitet wurden. Das von der Konstitutivgleichung beschriebene Kriechverhalten zeigt einen Übergang vom exponentiellen zum linearen Verhalten mit geringeren Spannungen. Um das Primärkriechen zu berücksichtigen wurde eine Funktion mit Dehnungsverfestigung verwendet. Das Versagen durch Kriechen wird hinreichend genau beschrieben, wenn in die Konstitutivgleichung zwei Schädigungsparameter und die dazugehörigen Evolutionsgleichungen eingeführt sind. Die Beschreibung des Verhaltens bezüglich der Langzeitfestigkeit basiert auf der Annahme des

Übergangs von duktilem zu sprödem Verhalten mit abnehmender Spannung. Die zwei Schädigungsparameter haben einen unterschiedlichen Charakter bei duktiler und spröder Schädigungsakkumulierung. Die Kriechkonstitutiv- und Evolutionsgleichungen sind durch einen Arrheniusansatz erweitert worden, so dass die Temperaturabhängigkeit berücksichtigt werden kann. Die vereinheitlichte mehraxiale Form des Kriech- und Schädigungsmodells für große Spannungsbereiche wird präsentiert. Ein neues Versagenskriterium, das sowohl die maximale Zugspannung als auch die effektive von Mises-Spannung beinhaltet, wird eingeführt.

Mehrere Kriech- und Schädigungsmodelle wurden für Analysen von Bauteilen mit FEM-basierte Software wie ABAQUS und ANSYS verwendet. Diese Modelle wurden mithilfe von benutzerspezifischen Subroutinen in die Software integriert. Um diese Subroutinen zu verifizieren, wurden mehrere Benchmark-Kriechprobleme aufgestellt und numerisch gelöst. Beispielsweise wird eine Simulation zur Abschätzung der Langzeitfestigkeit anhand des Gehäuses eines Quick-Stop Ventils einer Dampfturbine präsentiert. Die Ergebnisse belegen, dass der entwickelte Ansatz in der Lage ist, die Merkmale des Kriechschädigungsprozesses zu erfassen.

Contents

1 Basic Assumptions and Motivation   **1**
    1.1   Creep Phenomena   2
    1.2   Phenomenological Modeling   5
    1.3   Creep Deformation Mechanisms   9
    1.4   Creep and Damage Models 17
    1.5   Scope and Motivation 22

2 Conventional approach to creep-damage modeling   **27**
    2.1   Non-isothermal isotropic creep-damage model 28
         2.1.1   Uni-axial stress state 29
         2.1.2   Multi-axial stress state 31
    2.2   Anisotropic creep-damage models 32
         2.2.1   Model for anisotropic creep in a multi-pass weld metal . . . 32
         2.2.2   Murakami-Ohno creep model with damage induced anisotropy   40

3 Non-isothermal creep-damage model for a wide stress range   **43**
    3.1   Non-isothermal creep constitutive modeling 45
    3.2   Stress relaxation problem and primary creep modeling 53
         3.2.1   Stress relaxation problem 55
         3.2.2   Primary creep strain 60
    3.3   Non-isothermal long-term strength and tertiary creep modeling . . . 62
         3.3.1   Stress dependence 65
         3.3.2   Creep-damage coupling 67
         3.3.3   Temperature dependence 74
    3.4   Stress-dependent failure criterion 78

4	Creep estimations in structural analysis	85
	4.1 Application of FEM to creep-damage analysis	86
	4.2 Numerical benchmarks for the creep-damage modeling	88
	4.2.1 Purposes and applications of benchmarks	88
	4.2.2 Simply supported beam	90
	4.2.3 Pressurized thick cylinder	91
	4.3 Anisotropic creep of a pressurized T-piece pipe weldment	93
	4.3.1 Formulation of structural model	93
	4.3.2 Analysis of numerical results	96
	4.4 Creep-damage analysis of power plant components	99
	4.4.1 Previous experience in FEA	100
	4.4.2 Steam turbine quick-stop valve	102
5	**Conclusions and Outlook**	**111**
A	**Identification procedure of creep material parameters**	**115**
	A.1 Secondary creep stage	115
	A.2 Tertiary creep stage	117
	A.3 Primary creep stage	118
Bibliography		**121**

Chapter 1
Basic Assumptions and Motivation

Creep is the progressive time-dependent inelastic deformation under constant load and temperature. *Relaxation* is the time-dependent decrease of stress under the condition of constant deformation and temperature. For many structural materials, for example steel, both the creep and the relaxation can be observed above a certain critical temperature. The creep process is accompanied by many different slow microstructural rearrangements including dislocation movement, ageing of microstructure and grain-boundary cavitation.

The above definitions of creep and relaxation [152] are related to the case of uni-axial homogeneous stress states realized in standard material testing. But many responsible engineering structures and components are subjected to high temperature environment and complex mechanical loadings over a long time of operation. Examples include structural components of power generation plants, chemical facilities, heat engines and other high-temperature equipment. The life of these structures is usually limited by possible time-dependent creep processes. Under *creep in structures* one usually understands time-dependent changes of strain and stress states taking place in structural components as a consequence of external loading and temperature. Examples of these changes include progressive deformations, relaxation and redistribution of stresses, local reduction of material strength, etc. Furthermore, the strain and stress states are inhomogeneous and multi-axial in most cases. The aim of "creep modeling for structural analysis" is the development of methods to

1. Basic Assumptions and Motivation

predict time-dependent changes of stress and strain states in engineering structures up to the critical stage of creep rupture, see e.g. [34, 152].

Design procedures and residual life assessments for such responsible high-temperature structural components as pressure piping systems and vessels, rotors and turbine blades, casings of valves and turbines, etc. require the accounting for creep and damage processes. Chapter 1 is devoted to the discussion of basic features of materials creep and damage behavior, the overview of main creep-deformation mechanisms, the highlighting of main approaches to phenomenological creep modeling and the definition of the scope for this contribution.

1.1 Creep Phenomena

In design of engineering structures against creep it is necessary to seek the material and the shape which will carry the design loads, without failure, for the design life at the design temperature. The meaning of "failure" depends on the application. Following four main types of failure shown in Fig. 1.1 are distinguished in [24, 25]:

a) Displacement-limited applications, in which precise dimensions or small clearances must be maintained (as in discs and blades of turbines).

b) Buckling-limited applications subjected to compressive loads, in which creep strain can cause buckling failure (as in high pressure pipelines).

c) Stress-relaxation-limited applications in which an initial tension relaxes with time (as in the pretensioning of cables or bolts).

d) Rupture-limited applications, in which dimensional tolerance is relatively unimportant, but fracture must be avoided (as in steam turbine quick stop valve).

Design procedures and residual life assessments for such responsible high-temperature structural components as pressure piping systems and vessels, rotors and turbine blades, casings of valves and turbines, etc. require the accounting for creep and damage processes. The aim of creep modeling for structural analysis is the development of reliable methods and creep models to predict time-dependent

1.1. Creep Phenomena

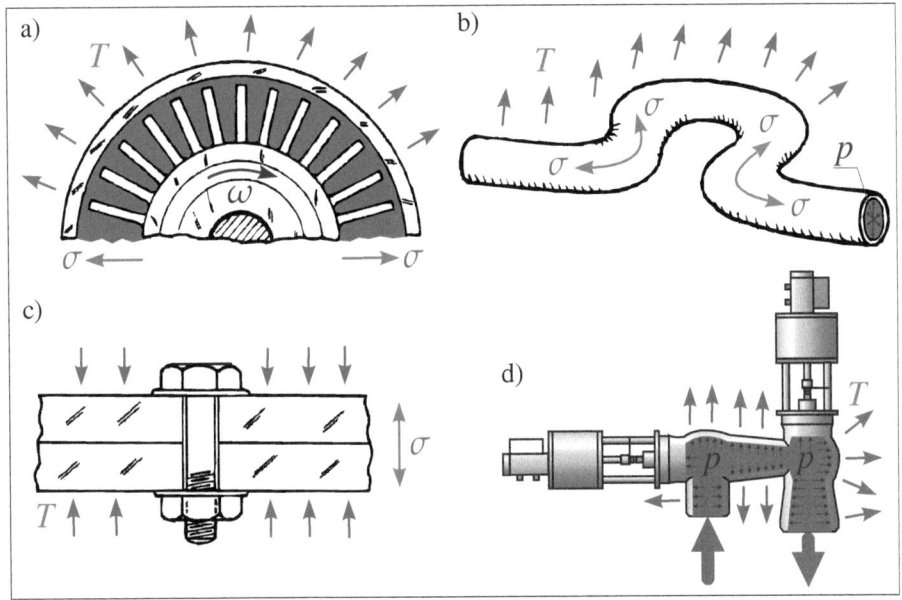

Figure 1.1: Creep is important in four classes of design [25]: a) displacement-limited, b) buckling-limited, c) relaxation-limited, d) failure-limited.

changes of stress and strain states in engineering structures up to the critical stage of creep rupture, see e.g. [34, 152].

Structural analysis under creep conditions requires a reliable constitutive model which reflects time dependent creep deformations and processes accompanying creep like hardening/recovery and damage. To tackle any of these we need constitutive equations which relate the creep strain rate $\dot{\varepsilon}^{cr}$ or time-to-failure t^* for the material to the stress σ and temperature T to which the material is exposed. The phenomenological approach to the development of a creep constitutive model is based on mathematical description of experimental creep curves obtained from uniaxial creep tests.

A standard cylindrical tension specimen is heated up to the temperature $T = (0.3 - 0.5)T_m$ (T_m is the melting temperature of the material) and loaded by a tensile force F. The value of the normal stress in the specimen σ_0 should be much less than the yield limit of the material σ_y. The instantaneous material response is therefore

1. Basic Assumptions and Motivation

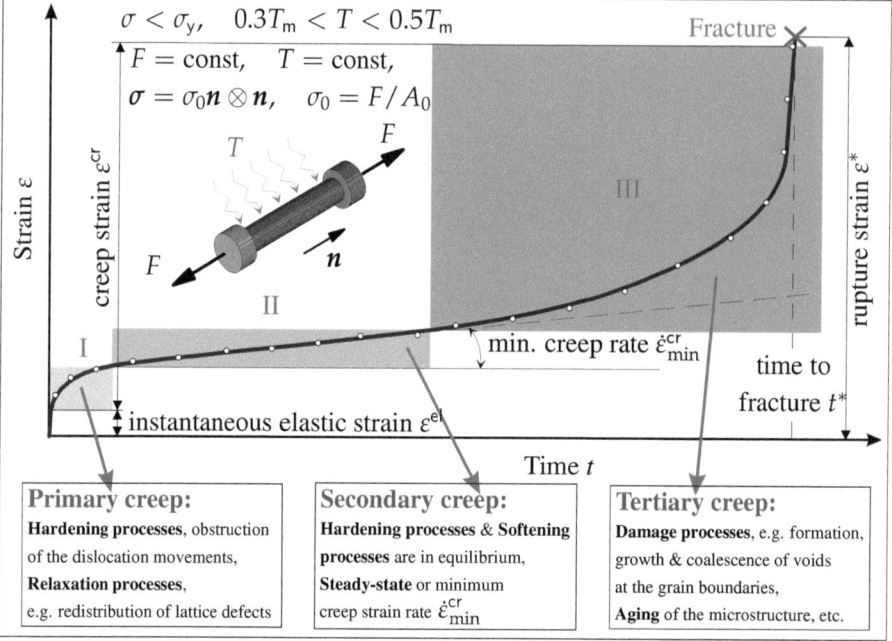

Figure 1.2: Strain vs. time curve under constant load F and temperature T (I – primary creep, II – secondary creep, III – tertiary creep), after [152].

elastic. The load and the temperature are kept constant during the test and the axial engineering strain ε is plotted versus time t. A typical creep curve for a metal is schematically shown in Fig. 1.2, the same examples of schematic representation of creep curve can be found in [24, 25, 145, 152, 181]. The instantaneous response can be characterized by the strain value ε^{el}. The time-dependent response is the slow increase of the strain ε with a variable rate. Following Andrade [57], three stages can be considered in a typical creep curve: the first stage (primary or reduced creep), the second stage (secondary or stationary creep) and the third stage (tertiary or accelerated creep). During the primary creep stage the creep rate decreases to a certain value (minimum creep rate $\dot{\varepsilon}^{cr}_{min}$). The secondary stage is characterized by the approximately constant creep rate. During the tertiary creep stage the strain rate increases. At the end of the tertiary stage creep rupture of the specimen occurs at time moment t^*.

A number of creep material properties can be deduced from the uniaxial creep curve. The most important of them are the duration of the stages, the value of minimum creep rate $\dot{\varepsilon}^{cr}_{min}$, the time to fracture t^* and the strain value before fracture ε^*. The shape of the creep curve is determined by several competing reactions [197] including:

1. Strain hardening;

2. Softening processes such as recovery, recrystallization, strain softening, and precipitate overaging;

3. Damaging processes, such as cavitation and cracking, and specimen necking.

Of these factors strain hardening tends to decrease the creep rate $\dot{\varepsilon}^{cr}$, whereas the other factors tend to increase the creep rate $\dot{\varepsilon}^{cr}$. The balance among these factors determines the shape of the creep curve. During primary creep the decreasing slope of the creep curve is attributed to strain hardening. Secondary-stage creep is explained in terms of a balance between strain hardening and the softening and damaging processes resulting in a nearly constant creep rate. The tertiary stage marks the onset of internal- or external-damage processes (item 3 in the preceding numbered list), which result in a decrease in the resistance to load or a significant increase in the net section stress. Coupled with the softening processes (item 2), the balance achieved in stage 2 is now offset, and a rapidly increasing tertiary stage of creep is reached.

1.2 Phenomenological Modeling

In general, let us consider an additive split of the uniaxial strains ε as follows, see e.g. [10]:

$$\varepsilon = \varepsilon^{el} + \varepsilon^{th} + \varepsilon^{inel} \qquad (1.2.1)$$

with ε^{el}, ε^{th}, ε^{inel} as the elastic, the thermal, and the inelastic strains, respectively. The third term can be split into a creep and a plastic part:

$$\varepsilon^{inel} = \varepsilon^{cr} + \varepsilon^{pl}. \qquad (1.2.2)$$

1. Basic Assumptions and Motivation

Below we neglect the thermal and the plastic strains considering only elastic and creep strains, as illustrated on Fig. 1.2. This simplification yields

$$\varepsilon = \varepsilon^{el} + \varepsilon^{cr}. \tag{1.2.3}$$

The phenomenological models of the creep theory are mostly based on constitutive relations of the following type

$$f(\varepsilon, \sigma, t, T) = 0, \tag{1.2.4}$$

where T denotes the temperature. At fixed temperature and prescribed stress history $\sigma(t)$ this equation determines the strain variation $\varepsilon(t)$ and vice versa [193]. The identification of Eq. (1.2.4) is connected with the performance of possible tests, for example, the creep or the relaxation test. For constant uni-axial stresses the creep law can be approximated by separating the stress, time, and temperature influences

$$\varepsilon^{cr} = f_1(\sigma)\, f_2(t)\, f_3(T), \tag{1.2.5}$$

where the representations for the stress function $f_1(\sigma)$, the time function $f_2(t)$, and the temperature function $f_3(T)$ are well-known from literature. A generalization for multi-axial states is possible, for instance, by analogy to the flow theory in plasticity. Below the attention will be focused on the stress and temperature functions. The examples of representations of stress, time and temperature functions can be found in [135, 192, 193].

The creep behavior can be divided into three stages as shown in Fig. 1.2. The first stage is connected with a hardening behavior, characterized by a decreasing creep strain rate. The second stage is the stationary creep with a constant creep strain rate (the creep strains are proportional with respect to the time). During this stage we consider an equilibrium between the hardening and the damage processes in the material. The last stage is the tertiary creep with an increasing creep strain rate, characterized by a dominant softening in the material. The main softening is realized by damage processes. Note that some materials show no tertiary creep, others have a very short primary creep period. In all cases we obtain rupture (failure) caused by creep as the final state of a creep curve.

The artificial modeling of the phenomenological creep behavior based on the division of the creep curve (creep strain versus time at constant stresses) allows the multi-axial creep description as follows. Let us introduce a set of equations, which contains three different types: an equation for the creep strain rate tensor influenced by hardening and/or damage and two sets of evolution equations for the hardening and the damage variables:

$$\dot{\varepsilon}^{cr} = \frac{\partial F(\sigma_{eq}, T; H_1, \ldots, H_n, \omega_1, \ldots, \omega_m)}{\partial \sigma},$$

$$\dot{H}_i = \dot{H}_i(\sigma_{eq}^H, T; H_1, \ldots, H_n, \omega_1, \ldots, \omega_m), \quad i = 1, \ldots, n, \quad (1.2.6)$$

$$\dot{\omega}_k = \dot{\omega}_k(\sigma_{eq}^\omega, T; H_1, \ldots, H_n, \omega_1, \ldots, \omega_m), \quad k = 1, \ldots, m.$$

In the notation (1.2.6) $\dot{\varepsilon}^{cr}$ is the creep strain rate tensor, σ is the stress tensor, F is the creep potential, H_i and ω_k are the hardening and damage variables, σ_{eq}^H, σ_{eq} and σ_{eq}^ω are the equivalent stresses which control the primary, secondary and tertiary creep. The proposed set of equations (1.2.6) can be used for classical and non-classical creep models, see e.g. [10].

It must be underlined that in addition to the problem how to formulate the set of creep equations (1.2.6) the identification problem for this set must be solved. In Fig. 1.3 the solution is schematically shown. The starting point is the identification of the creep equation for the secondary part. Since the creep is stress and temperature dependent this identification can be realized at fixed temperatures and for constant stress in a very easy way. If such an approach is not satisfying, we have to consider a more complex situation and use approaches presented, for example, in [81] and valid for a wide range of stresses.

The creep is influenced by hardening and damaging processes and based on the identified secondary creep equation. The equations for primary and tertiary creep can be established by extension of the secondary creep equations. The identification can be realized by analogy using additional experimental results. This approach is presented in several textbooks, for instance [124–126, 174, 191, 192].

The phenomenological approach to creep modeling generally includes 3 steps, see Fig. 1.3:

1. The first step of the phenomenological creep modelling is the formulation of empirical functions describing the sensitivity of the minimum creep rate $\dot{\varepsilon}_{min}^{cr}$ during the steady state to the stress level and the temperature. Hardening processes and softening processes are in equilibrium during secondary creep

1. Basic Assumptions and Motivation

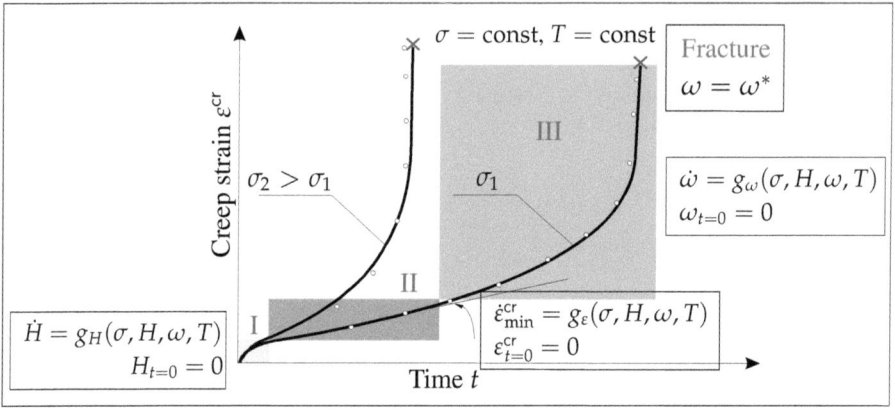

Figure 1.3: Identification of creep equations by uniaxial tests (after [10]).

stage. The steady-state (minimum) creep strain rate is defined by the constitutive equation $\dot{\varepsilon}^{cr}_{min} = g_\varepsilon(\sigma, T)$, which depends only on the stress σ and the temperature T. The power law and the hyperbolic functions of stress and Arrhenius functions for temperature dependence are mostly used in applications [145].

2. During the primary stage creep strain ε^{cr} is decelerated by hardening process (e.g. obstruction of the dislocation movements) and relaxation processes (e.g. redistribution of lattice defects). To characterize the hardening/recovery processes the constitutive equation $\dot{\varepsilon}^{cr}_{min} = g_\varepsilon(\sigma, T, H)$ is generalized by introduction of hardening internal state variable H and appropriate evolution equation $\dot{H} = g_H(\sigma, T, H)$.

3. During the tertiary stage creep strain ε^{cr} is accelerated by the damage process (e.g. nucleation, growth and coalescence of voids at the grain boundaries) and aging process (e.g. degradation of the material microstructure). To characterize the damage process the constitutive equation $\dot{\varepsilon}^{cr}_{min} = g_\varepsilon(\sigma, T, H, \omega)$ is generalized by introduction of the damage internal state variable ω and appropriate evolution equation $\dot{\omega} = g_\omega(\sigma, T, H, \omega)$.

The artificial modeling of the phenomenological creep behavior based on the division of the creep curve (creep strain ε^{cr} vs. time t at constant stress σ) allows the multi-axial creep description as follows. Let us introduce a set of equations, which contains three different types: an equation for the creep strain rate tensor $\dot{\boldsymbol{\varepsilon}}^{cr}$ influenced by hardening and/or damage and two sets of evolution equations for the hardening H and the damage ω variables. The complete system of equation, describing creep and accompanying processes, consisting of main constitutive equation for creep strain rate tensor $\dot{\boldsymbol{\varepsilon}}^{cr}$ and evolutionary equations for assumed internal state variables (H and ω) can be presented as follows:

$$\begin{cases} \dot{\boldsymbol{\varepsilon}}^{cr} = g_\varepsilon(\sigma, H, \omega, T), & \varepsilon^{cr}_{t=0} = 0 \quad \text{Constitutive Equation} \\ \dot{H} = g_H(\sigma, H, \omega, T), & H_{t=0} = 0 \quad \text{Evolution Equation for Hardening and Recovery} \\ \dot{\omega} = g_\omega(\sigma, H, \omega, T), & \omega_{t=0} = 0 \quad \text{Evolution Equation for Softening and Damage} \end{cases} \quad (1.2.7)$$

The emphasis in all cases of creep modeling is on correlating the macroscopic behavior with the underlying microscopic mechanisms. This requires that a variety of internal variables that describe the microstructural features that control the rate of deformation be considered. The internal state variables and the form of the creep potential can be chosen based on known mechanisms of creep deformation and damage evolution as well as possibilities of experimental measurement and engineering applications, e.g. [7, 13, 75].

1.3 Creep Deformation Mechanisms

Creep properties are generally determined by means of a test in which a constant uniaxial load or stress is applied to the specimen affected by an elevated temperature $T = (0.3 - 0.5)T_m$ and the resulting creep strain is recorded as a function of time. The influence of temperature and stress variation on the shape of creep curves are shown schematically in Fig. 1.4. After the instantaneous strain ε_0 a decelerating strain-rate stage (primary creep) leads to a steady minimum creep rate $\dot{\varepsilon}^{cr}_{min}$ (or secondary creep strain rate), which is finally followed by an accelerating stage (tertiary creep) that ends in fracture with a rupture strain ε^* at a rupture time t^*. It is observable from Fig. 1.4 that the value of the minimum creep strain rate $\dot{\varepsilon}^{cr}_{min}$ of creep

1. Basic Assumptions and Motivation

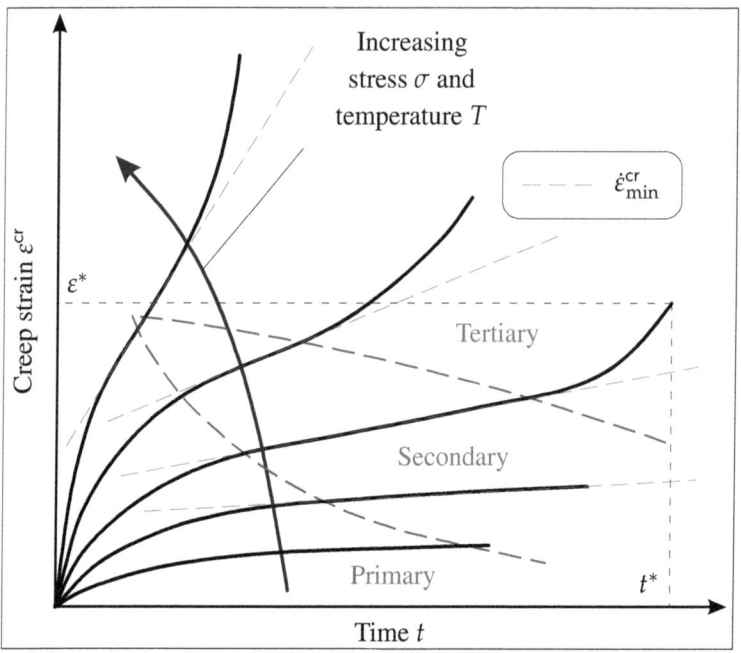

Figure 1.4: Schematic illustration of creep curve shapes, after [197].

curves with apparent secondary creep stage is rising with a growth of stress σ and temperature T values.

Materials can deform by dislocation plasticity or, if the temperature is high enough, by diffusional flow or power-law creep. If the stress σ and temperature T are too low for any of these, the deformation is elastic. To distinguish between the different mechanisms involved in creep damage, it is helpful to use a compact method of representation, partly developed by Graham and Walles [198] and later called by Frost and Ashby the "deformation-mechanism map" in [70]. Schematic illustration of typical map is shown in Fig. 1.5 in which the stress- and temperature-dependent regimes over which different types of creep processes dominate can be captured. Contours of constant strain rates are presented as functions of the normalized equivalent stress σ_{eq}/G and the homologous temperature T/T_m, where G is the shear modulus and T_m is the melting temperature. For a given combination of the stress σ and the temperature T, the map provides the dominant creep mechanism and

1.3. Creep Deformation Mechanisms

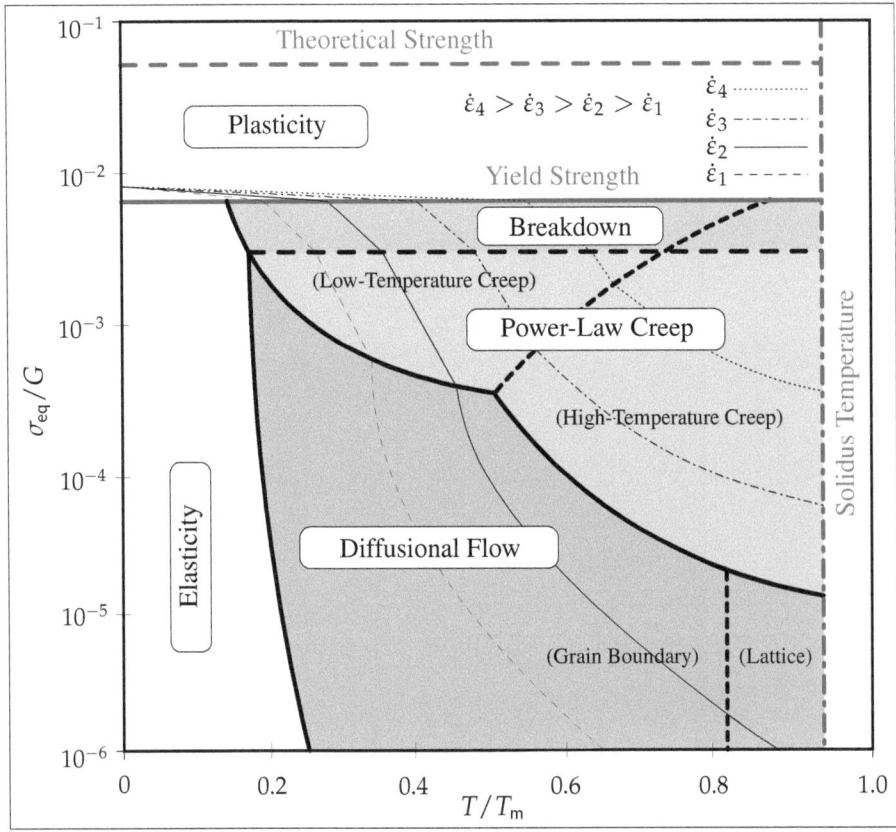

Figure 1.5: Schematic illustration of a typical *deformation-mechanisms map* for a metal material, after [70].

the strain rate $\dot{\varepsilon}$. It shows the range of stress σ and temperature T in which we expect to find each sort of deformation ε and the strain rate $\dot{\varepsilon}$ that any combination of them produces (the contours).

The first global overview of the *deformation mechanism maps* was provided by Frost and Ashby in [70]. Later a lot of examples for *deformation-mechanism map* of different materials were widely presented in literature, refer e.g. to [24, 25, 69, 145, 152, 166, 177, 181]. Diagrams like these are available for many metals and ceramics, and are a useful summary of creep behavior, helpful in selecting a material for high-

1. Basic Assumptions and Motivation

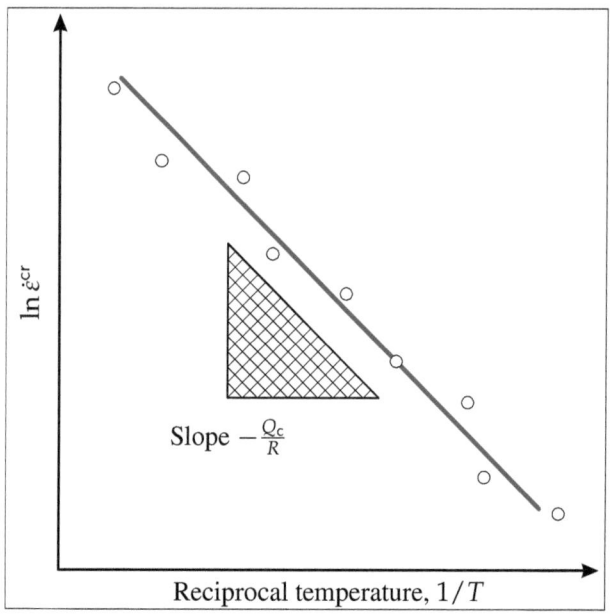

Figure 1.6: Temperature dependence of the minimum creep rate $\dot{\varepsilon}^{cr}$, after [24].

temperature applications.

Creep strain accumulation is a heat-activated process. An elementary deformation event gets additional energy from local thermal excitation. It is generally agreed that above 0.5 T_m (T_m is the melting temperature) the activation energy of steady-state deformation is close to the activation energy of self-diffusion. The correlation between the observed activation energy of creep Q_c and the energy of self-diffusion in the crystal lattice of metals Q_{sd} is illustrated in Fig. 1.7. For more than 20 metals an excellent correlation between both values has been documented in [127, 156]. Therefore, by plotting the natural logarithm (ln) of the minimum creep rate $\dot{\varepsilon}^{cr}$ against the reciprocal of the absolute temperature $(1/T)$ at constant stress $\sigma = $ const, as shows on Fig. 1.6 and proposed in [24], the correlation between $\dot{\varepsilon}^{cr}$ and T can be comprehensively expressed using the Arrhenius-type function as follows

$$\dot{\varepsilon}^{cr} \sim \exp\left(\frac{-Q_c}{RT}\right), \qquad (1.3.8)$$

where $R = 8.31 \ [\text{J} \cdot \text{mol}^{-1} \cdot \text{K}^{-1}]$ is the universal gas constant, and Q_c is called the

1.3. Creep Deformation Mechanisms

activation energy for creep with units of $[\text{J} \cdot \text{mol}^{-1}]$ and T is an absolute temperature with units of K.

At stresses σ and temperatures T of interest to the engineer, the following behavior proposed by Norton [157] and Bailey [27] is generally obeyed:

$$\dot{\varepsilon}^{cr} = A\,\sigma^n, \tag{1.3.9}$$

where A and n are stress-independent secondary creep constants. An exponential relationship, although not generally used, has also been proposed [185] to explain

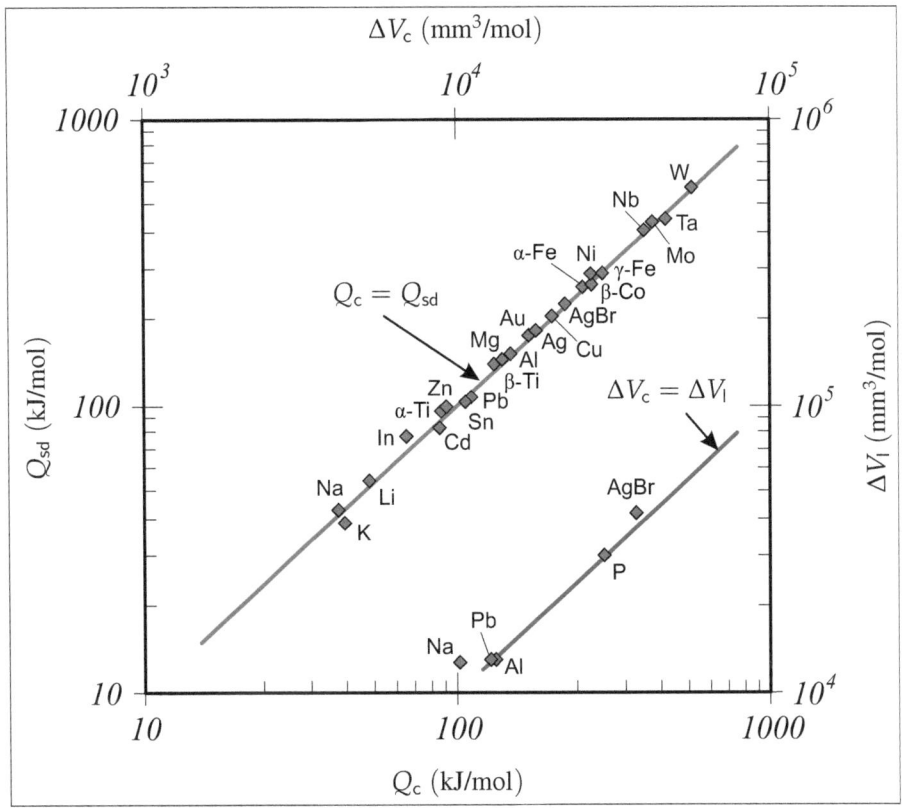

Figure 1.7: Comparison of the activation energy of creep Q_c and the activation energy of self-diffusion Q_{sd} for pure metals, after [156].

1. Basic Assumptions and Motivation

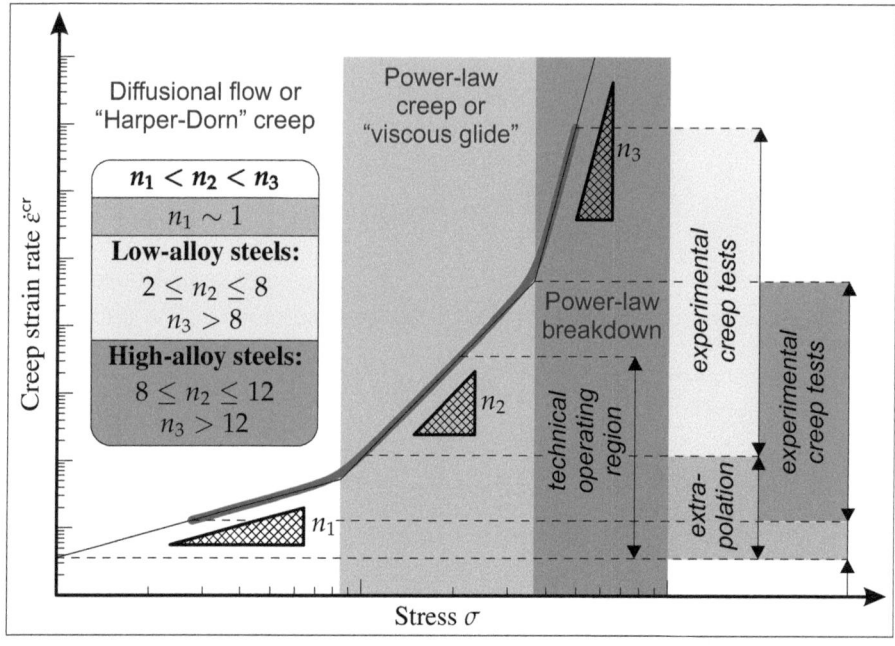

Figure 1.8: Schematic illustration of creep rate $\dot{\varepsilon}^{cr}$ vs. stress σ dependence, after [24, 25, 58, 59, 119, 203].

the behavior at very high stresses, as follows:

$$\dot{\varepsilon}^{cr} = A \, \exp\left(C \, \sigma\right), \tag{1.3.10}$$

where A and C are stress-independent constants.

Because creep is a thermally activated process, its temperature sensitivity would be expected to obey an Arrhenius-type expression (1.3.8), with a characteristic activation energy Q_c for the rate-controlling mechanism. Considering both the stress σ and temperature T dependencies of the creep strain rate $\dot{\varepsilon}^{cr}$, Eq. (1.3.9) can therefore be rewritten as [185]:

$$\dot{\varepsilon}^{cr} = A_0 \, \sigma^n \, \exp\left(\frac{-Q_c}{RT}\right), \tag{1.3.11}$$

where A_0 and n are stress-independent creep constants, and $R = 8.314 \, [\text{J} \cdot \text{K}^{-1} \cdot \text{mol}^{-1}]$ is the universal gas constant.

1.3. Creep Deformation Mechanisms

Table 1.1: Reported experimental values of secondary creep material parameters n_i and Q_i ($i = 2, 3$) for low-alloy heat-resistant steels, after [196, 197]

System of steel	Temperature, °C	Low-stress region n_2	Q_2, kJ/mole	High-stress region n_3	Q_3, kJ/mole	Reference
$1\frac{1}{4}Cr\text{-}\frac{1}{2}Mo$	510–620	4	400	10	625	[196]
$2\frac{1}{4}Cr\text{-}1Mo$	565	2.5	—	12	—	[41]
$Cr\text{-}Mo\text{-}V$	550–600	4.9	326	14.3	503	[68]
$Fe\text{-}V\text{-}C$	440–575	2.7	304	9.5	620	[52]
$Cr\text{-}Ni\text{-}Mn$	600–750	1.5–2	400-470	5.6	—	[23]

Although Eq. (1.3.11) suggests constant values for n and Q_c, experimental results on steels show both of these values to be variable with respect to stress σ and temperature T. The extended overviews of such experimental results are reported e.g. in [63, 196, 197]. Schematic illustration of creep strain rate $\dot{\varepsilon}^{cr}$ vs. stress σ dependence typical for the majority of advanced heat-resistant steels is shown on Fig. 1.8. Such a generalization of phenomenological approach to the description of creep behaviour and change of constant n has been proposed to use in [24,25,58,59,119,203].

The general approach illustrated on Fig. 1.8 shows the approximately stepped change in the value of creep constant n depending on the level of stress σ corresponding to the defined creep-deformation mechanism. For the all low-alloy and high-alloy heat resistant steels the stress exponent n from Eq. (1.3.11) is decreasing with the decrease of stress σ. But the change of creep constant n is stepped, because of the transition from the "power-law breakdown" at high stresses to the "power-law" creep mechanism at moderate stresses. This transition is laboratory well studied and reported for many low-alloy heat-resistant steels, see e.g. [23,41,52,68,196]. Table 1.1 shows the summary of reported experimental data for several low-alloy steels with n_3 and Q_3 as "power-law breakdown" creep constants and n_2 and Q_2 as creep constants corresponding to "power-law" mechanism. The reported values of stress exponent n can be generalized as $n_3 > 8$ for high stress levels and then gradually reducing to $2 \leq n_2 \leq 8$ at moderate stress levels, see Fig. 1.8.

With the development of advanced high-alloy steels, the activated creep-

1. Basic Assumptions and Motivation

Table 1.2: Reported experimental values of secondary creep material parameters n_i and Q_i ($i = 1, 2$) for high-alloy heat-resistant steels, after [196, 197]

System of steel	Tempera-ture, °C	Low-stress region		High-stress region		Reference
		n_1	$Q_1, \dfrac{kJ}{mole}$	n_2	$Q_2, \dfrac{kJ}{mole}$	
9Cr-1Mo-V-Nb	600–650	1	200	12	600	[106] [109] [189]
20Cr-25Ni-Nb	750	3–4.7	465–532	8–12	440–494	[138]
18Cr-10Ni-C	500–750	1	160	6	285	[31]
18Cr-12Ni-Mo	650–750	1	150-200	7	400-430	[110]
γ' austenitic	600	4.5	—	13	—	[195]

deformation mechanisms corresponding to the same stress levels as for low-alloy steels has changed. The technical operating region for high-alloy steels also includes the change of creep constant n with the decrease of stress σ. It is caused by the transition from the "power-law" creep at high stresses to the "linear" creep or diffusional flow mechanism at moderate stresses. This transition is not laboratory well studied, but never the less some experiments are reported for several high-alloy heat-resistant steels, see e.g. [31, 106, 109, 110, 138, 189, 195]. Table 1.2 shows the summary of reported experimental data for several high-alloy steels with n_2 and Q_2 as "power-law" creep constants and n_1 and Q_1 as creep constants corresponding to the "linear" creep mechanism. The reported values of stress exponent n can be generalized as $8 \leq n_2 \leq 12$ for high stress levels and then gradually reducing to $n_1 \sim 1$ at moderate stress levels, see Fig. 1.8.

Although many investigators report a distinct break in the curve presenting creep strain rate $\dot{\varepsilon}^{cr}$ vs. stress σ dependence, others view the value of n as continuously changing with stress σ and temperature T. While discussions continue regarding the natures of n and Q and the reasons for their variations, industrial practice has continued to ignore these controversies and to use a simple power-law Eq. (1.3.11) with discretely chosen values of n and Q. Because variations in n and Q are generally interrelated and self-compensating, no major discrepancies in the end results have

Table 1.3: Deformation mechanisms and corresponding response functions

Deformation mechanisms	Response functions	References
Power-law creep	$\dot{\varepsilon}_{cr} \propto \exp\left(-\frac{Q_c}{kT}\right) \sigma^n$	[27, 70, 157]
Diffusional flow	$\dot{\varepsilon}_{cr} \propto \exp\left(-\frac{Q_c}{kT}\right) \sigma$	[50, 82, 83, 91, 128, 144]
Linear + power-law	$\dot{\varepsilon}_{cr} \propto \exp\left(-\frac{Q_c}{kT}\right) \sinh(A\,\sigma)$	[61, 62]
Power-law breakdown	$\dot{\varepsilon}_{cr} \propto \exp\left(-\frac{Q_c}{kT}\right) \exp(C\,\sigma)$	[185]

been yet noted [197].

Actually, of all the parameters pertaining to the creep process, the most important for engineering applications are the minimum strain rate $\dot{\varepsilon}^{cr}$ and the time to rupture t^*. Specifically, their dependence on temperature T and applied stress σ are of the most interest to the engineer. This dependence varies with the applicable creep mechanism. A variety of mechanisms and equations have been proposed in the literature and have been reviewed elsewhere, e.g. [63, 73, 152, 166, 197]. Finally, the creep deformation mechanisms presented on the idealized deformation-mechanisms map (see Fig. 1.5) and corresponding response functions on stress σ and temperature T are listed in Table 1.3.

1.4 Creep and Damage Models

A typical creep behaviour of metals and alloys is accompanied by time-dependent creep deformations and damage processes induced by the nucleation and the growth of microscopic cracks and cavities. In order to characterise the evolution of the material damage as well as to describe the increase in creep strain rate during tertiary creep the continuum damage mechanics has been established and demonstrated to be a powerful approach, e.g. [88]. A lot of applications of creep continuum damage mechanics are related to the long-term predictions in thin-walled structures, e.g. pipes or pipe bends used in power and chemical plants. Here follows the short introduction into the continuum damage mechanics.

Damage accumulates in the form of internal cavities during creep. The damage first appears at the start of the *tertiary stage* of the creep curve and grows at an increasing rate thereafter. The shape of the *tertiary stage* of the creep curve (see Fig. 1.2) reflects this: as the cavities grow, the cross-section of the specimen decreases, and at constant load the stress σ goes up. Since $\dot{\varepsilon}^{cr} \propto \sigma^n$, the creep rate $\dot{\varepsilon}^{cr}$ goes up even faster than the stress σ does caused by creep damage, as illustrated on Fig. 1.9.

Isotropic damage models are generally formulated using the the concept of the

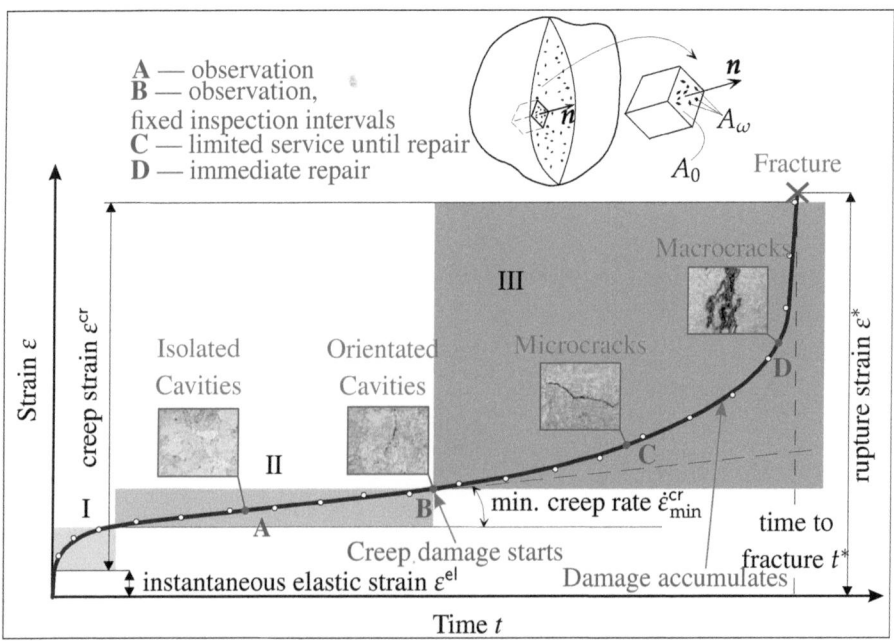

Figure 1.9: Evolution of damage caused by creep and corresponding service operations in a high-temperature component, after [154].

1.4. Creep and Damage Models

effective stress $\tilde{\sigma}$, see e.g. [34, 124, 152, 166]. In the uniaxial case this concept is formulated as follows. Previous studies in *continuum damage mechanics* starts with the concept of *continuity* ψ_n introduced by Kachanov [100]:

$$\psi_n = \frac{A_0 - A_\omega}{A_0}, \qquad (1.4.12)$$

where A_0 denotes the cross section of a uniaxial specimen, A_ω is the cross section area of cavities, and $\mathbf{n}$ is the normal vector to the cross-section, as illustrated on Fig. 1.9. Using the Eq. (1.4.12) a virgin state is characterized with $\psi_n = 1$, a fracture is characterized with $\psi_n = 0$, an isotropic damage is described with $\psi \equiv \psi_n$, and for the damaged state the *continuity* ψ_n lies in the range $1 \geq \psi_n \geq 0$.

Later the concept of *continuity* ψ_n was extended to the concepts of scalar state variable ω, e.i. isotropic *damage* parameter ω, which characterises a damage state of a material loaded by the stress σ. Specifying by ω the area fraction of cavities

$$\omega = A_\omega / A_0 \equiv 1 - \psi, \qquad (1.4.13)$$

one can introduce the *effective stress* or *net stress* $\tilde{\sigma}$ by dividing the applied force F to the *effective area* or *net area*

$$\tilde{A} = A_0 - A_\omega = A_0(1 - \omega). \qquad (1.4.14)$$

As a result the effective stress was defined by Rabotnov in [173] as follows:

$$\tilde{\sigma} = \frac{F}{\tilde{A}} = \frac{F}{A_0(1-\omega)} = \frac{\sigma}{(1-\omega)}. \qquad (1.4.15)$$

In [173] Rabotnov pointed out that the damage state variable ω "may be associated with the area fraction of cracks, but such an interpretation is connected with a rough scheme and is therefore not necessary". Rabotnov assumed that the creep rate is additionally dependent on the current damage state. The constitutive equation should have the form

$$\dot{\varepsilon}^{\text{cr}} = \dot{\varepsilon}^{\text{cr}}(\sigma, \omega). \qquad (1.4.16)$$

Furthermore, the damage processes can be reflected in the evolution equation

$$\dot{\omega} = \dot{\omega}(\sigma, \omega), \quad \omega|_{t=0} = 0, \quad \omega < \omega^*, \qquad (1.4.17)$$

where ω^* is the critical value of the damage parameter for which the material fails. With the power functions of stress and damage the constitutive equation may be formulated as follows

$$\dot\varepsilon^{cr} = \frac{A\,\sigma^n}{(1-\omega)^m}, \qquad (1.4.18)$$

accompanied by the damage evolution equation in the form

$$\dot\varepsilon^{cr} = \frac{B\,\sigma^k}{(1-\omega)^l}, \qquad (1.4.19)$$

where A, B, n, m, l, k are the material dependent creep parameters.

These Eqs (1.4.18) and (1.4.19) can be only applied to the case of constant temperature T. To generalize them to the non-isothermal conditions the material constants A and B should be replaced by the functions of temperature T. Assuming the Arrhenius-type temperature dependence (1.3.8) the following relations can be utilized [152]

$$A(T) = A_0 \exp\left(\frac{-Q_c}{RT}\right) \quad \text{and} \quad B(T) = B_0 \exp\left(\frac{-Q_d}{RT}\right), \qquad (1.4.20)$$

where Q_c and Q_d are the activation energies of creep and damage processes, respectively.

To identify the material constants in Eqs (1.4.18) - (1.4.20) experimental data of uni-axial creep up to rupture for certain stress and temperature ranges are required. The identification procedure is presented e.g. in [114]. If to limit to the case of fixed temperature and to assume $m = n$ in Eq. (1.4.18), then the uni-axial creep model takes the form [152]

$$\dot\varepsilon^{cr} = A\left(\frac{\sigma}{1-\omega}\right)^n \quad \text{and} \quad \dot\omega = \frac{B\,\sigma^k}{(1-\omega)^l} \qquad (1.4.21)$$

The model (1.4.21) was applied for the creep-damage description of transversely loaded shells and plates, presented e.g. in [13–15, 18, 19, 35, 51], and creep of several steels pipes and welded structures, e.g. in [42, 95, 151, 152]. The isotropic damage concept is suitable to characterise the creep-damage behaviour of some materials like steels and aluminium alloys [45] and is applicable for simple stress states typically realised in uniaxial creep tests [37, 143].

1.4. Creep and Damage Models

Dyson and McClean proposed in [61] the following modification of the Kachanov-Rabotnov creep-damage model, which fits all creep stages and it is mainly used for the creep modelling of low alloy ferritic steels and Ni-base alloys:

$$\dot{\varepsilon}^{cr} = \dot{\varepsilon}_0^{cr} (1 + D_d) \exp\left(\frac{-Q}{RT}\right) \sinh\left(\frac{\sigma(1-H)}{\sigma_0(1-D_p)(1-\omega)}\right), \quad (1.4.22)$$

where $\dot{\varepsilon}^{cr}$ and $\dot{\varepsilon}_0^{cr}$ are the equivalent minimum creep strain rate and the reference creep strain rate; σ and σ_0 are the equivalent stress and the reference stress, respectively; T is the temperature; Q is the creep activation energy; H, D_d, D_p, ω are internal state variables, defined by evolution equations. Notably, H is the hardening parameter, D_d is the damage parameter caused by multiplication of mobile dislocations, D_p is the damage parameter caused by particle coarsening and ω is the damage parameter caused by the cavity nucleation and growth.

There are unlimited possibilities to extend the constitutive equations. As one example can serve the model [167], in which the additional number of creep parameters, determining damage mechanisms and corresponding temperature dependencies, are introduced. A set of constitutive equations have been derived with an associated set of temperature dependencies, which describe the accumulation of intergranular cavitation, the coarsening of carbide precipitates and the influence of these mechanisms on the effective creep strain rate

$$\begin{aligned}
\dot{\varepsilon}_e &= A \sinh\left[\frac{B\sigma_e(1-H)}{(1-\Phi)(1-\omega)}\right], & \dot{\omega} &= CN\dot{\varepsilon}_e \left(\frac{\sigma_1}{\sigma_e}\right)^\nu, \\
\dot{H} &= \left(\frac{h\dot{\varepsilon}_e}{\sigma_e}\right)\left(1 - \frac{H}{H^*}\right), & \dot{\Phi} &= \left(\frac{K_c}{3}\right)(1-\Phi)^4, \\
A &= A_0 B \exp\left(-\frac{Q_A}{RT}\right), & B &= B_0 \exp\left(-\frac{Q_B}{RT}\right), \\
K_c &= \left(\frac{K_{c0}}{B^3}\right)\exp\left(-\frac{Q_{K_C}}{RT}\right), & C &= C_0 \exp\left(-\frac{Q_C}{RT}\right),
\end{aligned} \quad (1.4.23)$$

where $N = 1$ for $\sigma_1 > 0$ and $N = 0$ for $\sigma_1 \leq 0$; A_0, B_0, C_0, K_{c0}, h, H^*, Q_A, Q_B, Q_C and Q_{K_C} are material constants to be determined from uniaxial creep data over a range of stresses and of temperatures; the constant ν is determined from multi-axial creep rupture data [167].

1. Basic Assumptions and Motivation

1.5 Scope and Motivation

In recent years a lot of industrial and scientific organizations work intensively for improved modeling and experimental investigation of creep-damage behavior of heat-resistant materials used in high-temperature equipment components of power-generation plants, chemical facilities, heat engines, etc. Among them are European Creep Collaborative Committee, see e.g. [92, 139], Forschungszentrum Karlsruhe (Germany), see e.g. [117, 178, 179], Institute of Physics of Materials Brno (Czech Republic), see e.g. [105–110, 165, 187–189], Materialprüfungsanstalt Stuttgart (Germany), see e.g. [74, 104, 133, 199], Oak Ridge National Laboratory (USA), see e.g. [111, 137], National Research Institute for Metals (Japan), see e.g. [5, 6, 118, 162–164, 202], National Aeronautics and Space Administration (USA), see e.g. [175, 176], etc. At present, a large number of creep models, which are able to describe uniaxial creep curves for certain stress and temperature ranges, have been developed and practically applied for creep estimations and life-time assessments. Only few of them are found applicable to the FEM-based creep modeling for structural analysis and life-time assessment under multi-axial stress states, because of their different initial designations and corresponding mathematical and mechanical limitations. Therefore, all the basic approaches to the description of creep behavior can be conventionally systematized on four main groups as proposed in [147, 152]: 1) *empirical models*, 2) *materials science models*, 3) *micro-mechanical models*, 4) *continuum mechanics models*.

Within the frames and objectives of this work the *continuum mechanics models* are of the most interest. The objective of *continuum mechanics modeling* is to investigate creep in idealized three-dimensional solids. The idealization is related to the hypothesis of a continuum, e.g. refer to [84]. The approach is based on balance equations and assumptions regarding the kinematics of deformation and motion. Creep behavior is described by means of constitutive equations which relate deformation processes and stresses. Details of topological changes of microstructure like subgrain size or mean radius of carbide precipitates are not considered. The processes associated with these changes like hardening, recovery, ageing and damage can be taken into account by means of hidden or internal state variables and corresponding evolution equations, see e.g. [34, 126, 174, 191]. Creep constitutive equations with internal state variables can be applied to structural analysis. Various models and

1.5. Scope and Motivation

methods recently developed within the mechanics of structures can be extended to the solution of creep problems. Examples are theories of rods, plates and shells as well as direct variational methods, e.g. [11, 34, 40, 135, 152, 168, 192]. Numerical solutions by the finite element method combined with various time step integration techniques allow to simulate time dependent structural behavior up to critical state of failure. Examples of recent studies include circumferentially notched bars [86], pipe weldments [90] and thin-walled tubes [117]. In these investigations qualitative agreements between the theory and experiments carried out on model structures have been established. Constitutive equations with internal state variables have been found to be mostly suited for the creep analysis of structures [90]. However, it should be noted that this approach requires numerous experimental data of creep for structural materials over a wide range of stress and temperature as well as different complex stress states.

This thesis is a contribution to the continuum mechanics modeling of creep and damage with the aim of structural analysis of industrial structures. This type of modeling is related to both fields of "creep mechanics" [34, 160] and "continuum damage mechanics" [88] and requires the following steps [40, 93, 147, 152, 159]:

- formulation of a constitutive model including creep constitutive equation and evolution equations for internal state variables (e.g. damage, softening, hardening, etc.) to reflect basic features of creep behavior of a structural material under multi-axial stress states,

- identification of creep material parameters in constitutive and evolution equations based on experimental data of creep and long-term strength at several temperatures,

- development of geometrical model of analysed structure or application of a structural mechanics model by taking into account creep processes and stress state effects,

- formulation of an initial-boundary value problem based on the creep constitutive and structural mechanics models with initial and boundary conditions,

- application of finite element method or development of numerical solution procedures,

1. Basic Assumptions and Motivation

- verification of obtained results as well as the numerical methods and algorithms.

The principal aim of this work is the development of a comprehensive creep-damage constitutive model based on continuum mechanics approach, then the following significant problems and questions of constitutive modeling must be taken into account:

- *Applicability to wide ranges of temperature and stress.* The constitutive and evolution equations are formulated as phenomenological dependencies on stress and temperature. According to experimental studies many materials for high-temperature applications exhibit a stress and temperature ranges dependent creep behavior. The European and Japanese material research results show that, especially for new steels in the class of 9-12%Cr, long-term creep-rupture testing of > 50 000 h under low stresses is required to determine the real alloy behaviour, e.g. refer to [101, 118]. Otherwise a very dangerous overestimation of the material stability can result as shown in different publications if comparing long-term extrapolations on the basis of either short or long term tests. Thus, the response functions of the applied stress and temperature in constitutive and evolutions equations should extrapolate the laboratory creep and rupture data usually obtained under increased stress and temperature to the in-service loading conditions relevant for industrial applications.

- *Ability to model tertiary creep stage.* The aim of creep modeling is to reflect basic features of creep in structures including the development of inelastic deformations, relaxation and redistribution of stresses as well as the local reduction of material strength. A model should be able to account for material degradation processes in order to predict long-term structural behavior, i.e. time-to-fracture, and to analyze critical zones of creep failure.

- *Minimum quantity of creep material parameters.* Conventional constitutive models contain the large number of parameters which must be identified by the way of complex, long and expensive experiments. Such situation results to high necessity for development of models with a small number of parameters at preservation accuracy, sufficient for practical purposes. Effective way of creation of such models is the combination of approaches of continuum mechanics with qualitative conclusions of physics of creep of metals.

1.5. Scope and Motivation

- *Simplicity of creep parameters identification.* Depending on the choice of creep-damage model the main problem is the creep parameter identification for the corresponding material. The significant scatter of properties of creep demands a lot number of experiments for statistical processing of experimental results. For parameter identification even simple constitutive models require a considerable number of experiments both under uni-axial and multi-axial loading. Thus, the creep material parameters for a selected constitutive model must be derived from the standard procedures for creep uni-axial testing.

- *Compatibility with commercial FEM-based software.* Commercial finite element codes like ABAQUS and ANSYS were developed to solve various problems in solid mechanics. In application to the creep analysis one should take into account that a general purpose constitutive equation which allows to reflect the whole set of creep and damage processes in structural materials over a wide range of loading and temperature conditions is not available at present in commercial FEM-based software. Therefore, a specific constitutive model with selected internal state variables, special types of stress and temperature functions as well as material constants identified from available experimental data should be incorporated into the commercial finite element code by writing a user-defined material subroutine [152]. Furthermore, the constitutive model must be compatible with standard programming languages (e.g. FORTRAN, C++, PYTHON) and easy to be translated into programming code.

Therefore, within the framework of the dissertation a comprehensive non-isothermal creep-damage model based on continuum mechanics approach and applicable to a wide stress range have to be developed. The model must rely on the material science assumptions about various creep-deformation behavior in a wide stress range and experimental observations showing the changes of material microstructure caused by long-term thermal exposure and creep deformations. In addition, it must contain a minimum of creep material parameters, which are easy identified from the standard procedures for creep and rupture uni-axial testing. For the purpose of effective application to numerical structural analysis in FEM-based software the model should be presented in the form of a user-defined material subroutine coding.

Chapter 2

Conventional approach to creep-damage modeling

In Chapter 1 we discussed theoretical approaches to the constitutive modeling of creep behavior. Chapter 2 presents the conventional approaches to creep-damage modeling, which are applied to several engineering materials. The models include specific forms of the constitutive equation for the creep rate tensor and evolution equations for internal state variables. In addition, constitutive functions of stress and temperature are specified. In order to find a set of material constants creep tests under constant load and temperature leading to a homogeneous stress state are required. The majority of available experimental data is presented as creep strain versus time curves from standard uni-axial tests. Based on these experimental curves the creep material parameters are identified.

Section 2.1 is devoted to description of the conventional isotropic Kachanov-Rabotnov-Hayhurst creep-damage model based on continuum damage mechanics, see e.g. [88, 100, 173]. The model is extended to the case of variable temperature and strain hardening consideration [114]. Both the creep and the damage rates are assumed temperature dependent. A technique for the identification of material creep parameters for the model based on the available family of experimental creep curves is presented in Appendix A. The model was applied to the numerical long-term strength analysis and life-time assessment under the creep conditions of several typical power-generation plant components [79, 114, 129–132].

The objective of Sect. 2.2.1 to develop a model for anisotropic creep behavior in

2. Conventional approach to creep-damage modeling

a weld metal produced by multi-pass welding. The anisotropy of creep properties is caused the complex directional microstructure of weld metal induced by the heat affect [94]. The structural analysis of a welded joint requires a constitutive equation of creep for the weld metal under multi-axial stress states. For this purpose we apply the approaches to modeling of creep for initially anisotropic materials. The outcome is the a creep constitutive equation for the strain rate tensor describing secondary and tertiary creep behavior. It is accompanied by the two damage evolution equations describing the different damage accumulation in the longitudinal direction and in the transverse plane of isotropy. The material constants are identified according to the experimental data presented in the literature [94]. The model was verified applying it to the numerical long-term strength analysis of a typical T-piece pipe weldment [80].

Since the nature of damage phenomenon is generally anisotropic, the isotropic material damage is just a simplified case of the damage anisotropy. Then it is necessary to highlight in Sect. 2.2.2 the Murakami-Ohno creep-damage model with damage induced anisotropy. The anisotropic creep behaviour induced by damage is characterised by introducing a tensor-valued internal state variable. We should discuss the anisotropic damage concept proposed by Murakami and Ohno [142] in order to conclude about the influence of damage induced anisotropy on the long-term predictions in high-temperature engineering applications, e.g. [13, 79].

2.1 Non-isothermal isotropic creep-damage model

The conventional isotropic creep-damage model by Kachanov [100], Rabotnov [173] and Hayhurst [88] is based on the continuum damage mechanics. It contains the power-law stress response function and a scalar damage parameter. The conventional model is extended to the case of variable temperature and strain hardening consideration. Both the creep and the damage rates are assumed temperature dependent using the Arrhenius-type functions. The constitutive model is able to describe the primary, secondary and tertiary stages of creep behavior. Both the uni-axial and the multi-axial forms of the model accompanied with a technique for the identification of material creep parameters based on the available family of experimental creep curves at various temperatures and wide stress range are presented below.

2.1. Non-isothermal isotropic creep-damage model

2.1.1 Uni-axial stress state

The Kachanov-Rabotnov-Hayhurst model and physical mechanisms of creep for typical heat-resistant steels build the basis for the here suggested non-isothermal creep-damage model. The primary creep is characterized by the introduction of the following strain hardening function:

$$H(\varepsilon^{cr}) = 1 + C \exp\left(-\frac{\varepsilon^{cr}}{k}\right). \tag{2.1.1}$$

In order to reflect various influences of temperature on the diffusional creep and the cross slip dislocation two different functional dependences are introduced in the constitutive creep strain rate equation and in the evolution equation determining the damage rate. For the description of temperature dependence the Arrhenius functions [167] are introduced:

$$A(T) = A \exp\left(\frac{-Q_\alpha}{RT}\right) \quad \text{and} \quad B(T) = B \exp\left(\frac{-Q_\beta}{RT}\right). \tag{2.1.2}$$

The uni-axial form of creep-damage equations considering strain hardening for variable temperature field is given in the following form

$$\frac{d\varepsilon^{cr}}{dt} = A(T) H(\varepsilon^{cr}) \left[\frac{\sigma}{1-\omega}\right]^n, \tag{2.1.3}$$

$$\frac{d\omega}{dt} = B(T) \frac{\sigma^m}{(1-\omega)^l}. \tag{2.1.4}$$

In Eqs (2.1.1) - (2.1.4) ε^{cr} represents the creep strain; t denotes time; σ is the uni-axial stress; Q_α and Q_β are energies of activation; T is the absolute temperature; A, B, C, n, m, k, l are the material constants; ω is the scalar damage parameter ($0 \leq \omega \leq \omega^*$), where ω^* is the critical value of damage corresponding to the time of rupture t^*.

Instead of three constants including the energies of activation of creep and damage processes and universal gas constant the following two constants are introduced:

$$h = \frac{Q_\alpha}{R} \quad \text{and} \quad p = \frac{Q_\beta}{R}. \tag{2.1.5}$$

In the general case, the values of thermal energies of activation for creep process Q_α and for damage process Q_β of the typical heat-resistant steel are different.

2. Conventional approach to creep-damage modeling

Together with the other material constants, they are estimated from the set of experimental creep curves within a wide range of temperatures and stresses.

The time integration of the damage evolution equation (2.1.4) under assumption of the fixed stress and temperature provides the function $w(t)$ as follows:

$$w(t) = 1 - \left[1 - (l+1)\ B\ \exp\left(\frac{-p}{T}\right)\sigma^m\ t\right]^{\frac{1}{l+1}}. \qquad (2.1.6)$$

Therefore, time-to-rupture t^* can be defined with assumption that $w = 1$ as follows:

$$t^* = \frac{1}{\left[(l+1)\ B\ \exp\left(\frac{-p}{T}\right)\sigma^m\right]}. \qquad (2.1.7)$$

By taking into account the time-dependent function $w(t)$ in the form of Eq. (2.1.6), the creep constitutive equation (2.1.3) is integrated by time as follows

$$\varepsilon^{cr}(t) = k\ \ln\left[(1+C)\ \exp\left(\frac{\zeta(t)}{k}\right) - C\right], \qquad (2.1.8)$$

where the auxiliary time-dependent function $\zeta(t)$ is defined in the following form:

$$\zeta(t) = \frac{A\ \exp\left(\frac{p-h}{T}\right)\sigma^{n-m}}{B\ (n-l-1)} \cdot \left\{\left[1 - (l+1)\ B\ \exp\left(-\frac{p}{T}\right)\sigma^m\ t\right]^{\frac{l-n+1}{l+1}} - 1\right\}. \qquad (2.1.9)$$

The procedure of material creep constants identification under the constant temperature is described in [77]. For the identification of values for the activation energies of creep-damage processes, i.e. creep constants h and p, it is necessary to have experiment data under at least two differents values of temperature T. If the experimental data is available for more than two temperature T values, than the identification procedure cited in [114] must be applied. The typical creep material parameters identification procedure for the uniaxial form (2.1.3) - (2.1.4) of conventional non-isothermal creep-damage model is highlighted in Appendix A.

For the creep rupture time assessment, which is principally defined by secondary and tertiary creep stages, it is possible to apply the simplified version of creep-damage model by neglecting the primary creep stage, i.e. by setting $H(\varepsilon^{cr}) = 1$.

Further simplification can be made by equating the tertiary creep constants $m = 1$. Such a model contains only 6 material constants, which can be easily determined from experimental data.

2.1.2 Multi-axial stress state

The internal material state variables and the form of the creep potential of the constitutive model for isotropic creep behaviour can be chosen based on known mechanisms of creep deformation and damage evolution [7]. According to known deformation mechanisms the primary and secondary creep rates are dominantly controlled by the von Mises effective stress. The tertiary creep stage, accelerated by damage, is additionally influenced by the kind of the stress state.

The conventional isotropic creep-damage model by Kachanov-Rabotnov-Hayhurst is based on the power-law stress function and a scalar damage parameter, with a constitutive equation

$$\dot{\boldsymbol{\varepsilon}}^{cr} = \frac{3}{2} \frac{\dot{\varepsilon}_{eq}^{cr}}{\sigma_{vM}} \boldsymbol{s}, \qquad (2.1.10)$$

where the equivalent creep strain rate have following form

$$\dot{\varepsilon}_{eq}^{cr} = A \, \exp\left(-\frac{h}{T}\right) \left[1 + C \, \exp\left(-\frac{\varepsilon_{eq}^{cr}}{k}\right)\right] \left(\frac{\sigma_{vM}}{1-\omega}\right)^n, \qquad (2.1.11)$$

and the evolutional equation for a scalar damage parameter is formulated as follows

$$\dot{\omega} = B \, \exp\left(-\frac{p}{T}\right) \frac{(\langle \sigma_{eq}^{\omega} \rangle)^m}{(1-\omega)^l}. \qquad (2.1.12)$$

In notations (2.1.10) - (2.1.12) $\dot{\boldsymbol{\varepsilon}}^{cr}$ represents the creep strain rate tensor; $\sigma_{vM} = \left[\frac{3}{2} \boldsymbol{s} \cdot \cdot \boldsymbol{s}\right]^{\frac{1}{2}}$ is the effective von Mises stress; $\boldsymbol{s}$ is the stress deviator; A, B, n, m, l are creep material parameters; ω is an isotropic damage parameter ($0 \leq \omega \leq \omega^*$), σ_{eq}^{ω} is the damage equivalent stress, used in the form proposed by Leckie and Hayhurst in [121]

$$\sigma_{eq}^{\omega} = \alpha \sigma_I + (1-\alpha) \, \sigma_{vM}, \qquad (2.1.13)$$

where σ_I is a maximum principal stress, α is a weighting factor considering the influence of damage mechanisms (σ_I-controlled or σ_{vM}-controlled).

2. Conventional approach to creep-damage modeling

The creep-damage model (2.1.10) - (2.1.12) fulfils the condition of incompressibility. Furthermore, the damage evolution is assumed only for the positive equivalent stress

$$\langle \sigma_{eq}^{\omega} \rangle = \sigma_{eq}^{\omega} \quad \text{for} \quad \sigma_{eq}^{\omega} > 0 \quad \text{and} \quad \langle \sigma_{eq}^{\omega} \rangle = 0 \quad \text{for} \quad \sigma_{eq}^{\omega} \leq 0. \tag{2.1.14}$$

2.2 Anisotropic creep-damage models

2.2.1 Model for anisotropic creep in a multi-pass weld metal

The lifetime to fracture of high-temperature components of fossil power plants and chemical facilities structures is defined by irreversible processes of creep and damage. The most dangerous and the probable place of rupture in welded structures of pipelines or high-pressure vessels is the weldment zone. The reason of that fact is a complex and non-uniform microstructure of the material in the weldment zone caused by the manufacturing process of multi-pass welding. It is possible to define at least three main zones: weld metal, adjacent heat-affected zone and parent material of a welded structure (see Fig. 2.1), as described in [95, 152].

The principal purposes of creep-damage modeling for welded structures are:

- the long-term prediction of stress redistribution in a local weldment zone having complex geometry as a result of creep strain,

- an estimation of probable zones with critical damage accumulation that could lead to fracture and crack initiation.

Particularly, creep modeling allows to predict zones of unexpected fracture and for the purpose of inspection of some structural components during the the service and to estimate the residual lifetime of a structure. Description and methodology of in-service inspections for high-temperature power and chemical plant components can be found e.g. in [43].

Analysis of literature

One of the possibilities to model creep process in welded structures is to use the concepts of continuum damage mechanics which provide constitutive equation for

2.2. Anisotropic creep-damage models

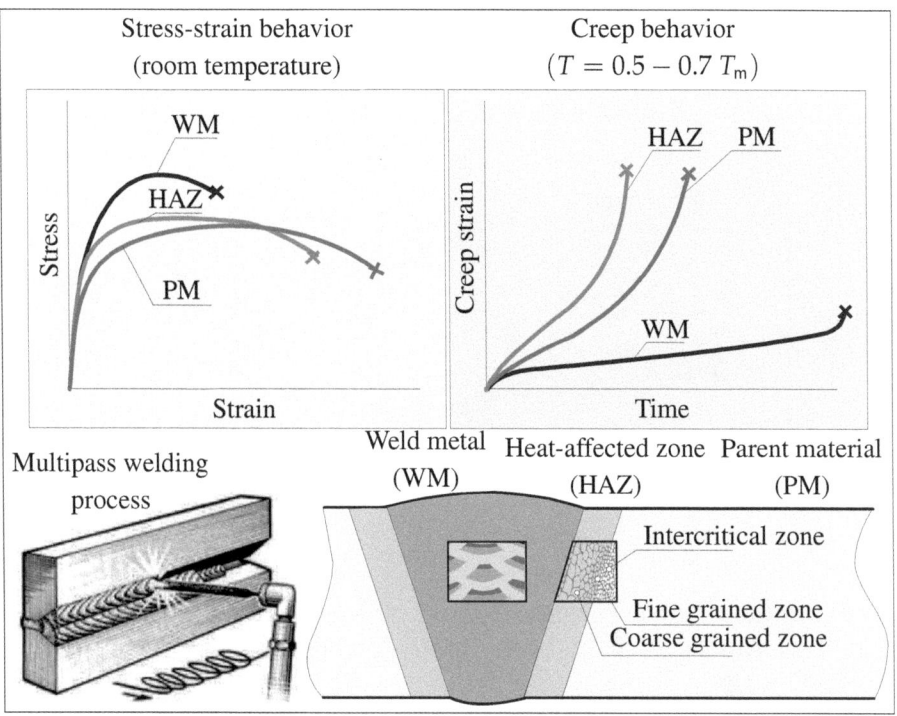

Figure 2.1: Typical microstructure of welded joint and material behavior, after [151, 152].

the creep strain rate tensor, damage evolution equation and evolution equations for other material state variables, e.g. hardening, softening, etc. These equations are complemented with non-linear initial-boundary problem to analyze the long-term strength of structures, see e.g. [13, 88]. The practical experience and approaches to creep modeling in welded structures using FEM during the last 10 years can be found in [90, 95, 186]. According to that approaches the weldment volume was divided on three zones considering different material behavior under the creep conditions (see Fig. 2.2).

The results of experimental observations [94, 95] generally show significant differences in the creep properties in a weldment zones with different microstructure. Thus, HAZ has higher creep strain rate and less time to rupture comparing to the

2. Conventional approach to creep-damage modeling

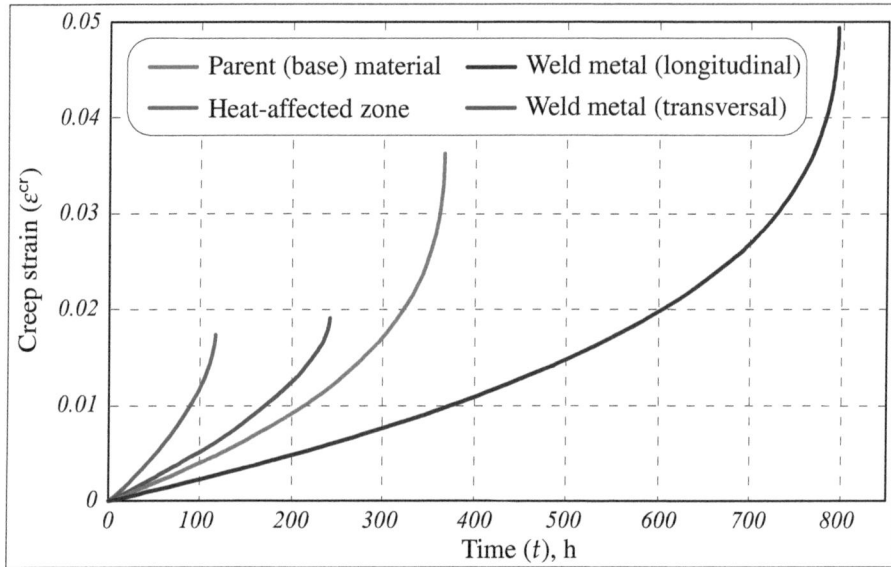

Figure 2.2: Creep curves of uniaxial specimen made of high-alloy heat-resistant steel P91 (9CrMoNbV) at temperature 650°C and tensile stress 100 MPa, after [80].

same characteristics of base (parent) material and weld metal. In general, results of finite-element creep modeling predicted the critical damage accumulation and further rupture with crack initiation in the fine-grained and intercritical HAZ. Such a type of fracture agrees with the experiments, and the weldment crack location is of type IV (see Fig. 2.3 and Table 2.1) due to the classification of for damage types in weldments [47]. However, failures due cracks within the weld metal have been encountered in practice, refer to [94]. These cracks have types I and II in the classification scheme for damage types in weldments [47], see Fig. 2.3 and Table 2.1.

Creep experimental data

Modeling of weld metal creep behavior based on experimental data [94] for the steel P91 (9CrMoNbV). Experimental observations demonstrated significant anisotropy of weld metal elastic and creep properties (see Fig. 2.4). The reason of anisotropic creep properties of weld metal in a welding joint is an orientation and microstruc-

2.2. Anisotropic creep-damage models

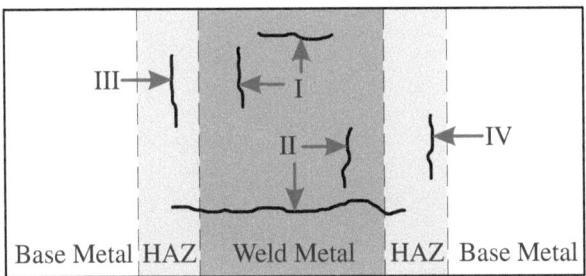

Figure 2.3: Classification scheme for damage types in weldments, after [47].

Table 2.1: Definition of weldment cracks, after [47].

Designation	Location of crack
Type I	in weld metal
Type II	in weld metal and adjacent HAZ
Type III	in coarse-grained HAZ
Type IV	in intercritical HAZ

tural inhomogeneity of weld metal caused by the manufacturing process of multi-pass welding (see Fig. 2.1). Experimental data [94] showed that creep curves are significantly different for tests of uniaxial specimens carved from weld metal along the welding direction and in transverse plane to it (see Fig. 2.4).

Previous experimental works and modeling approaches studying the creep of weldments (e.g. [94, 95, 104]) do not take into account the qualitative anisotropic creep behavior of weld metal in a multi-pass weld joint. Within the frames of the research work published in [80] experimental creep and damage data of multi-pass weldment of a high-alloy steel 9CrMoNbV [94], including the inhomogeneity of the microstructure in the heat affected zone, in the base material and in the weld metal of welding joint have been investigated and processed. For the purpose of adequate creep behavior modeling and long-term strength analysis of the welding seam using continuum damage mechanics approach the transversally-isotropic creep-damage

2. Conventional approach to creep-damage modeling

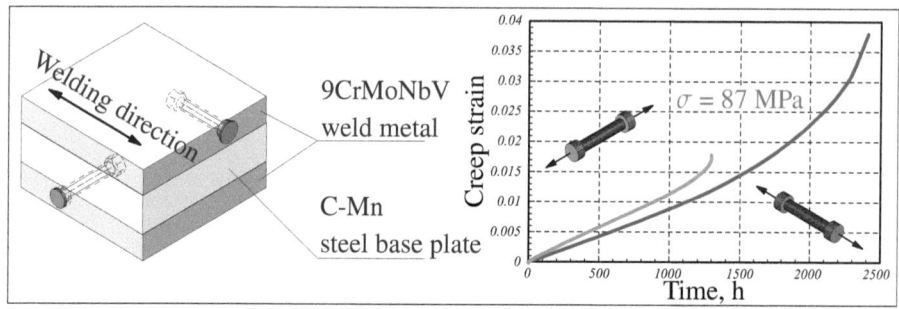

Figure 2.4: Experimental creep curves of P91 (9CrMoNbV) weld metal at temperature 650°C and tensile stress $\sigma = 87$ MPa, after [94].

model have been developed.

Transversely isotropic creep-damage model for weld metal

Long-term strength analysis of welded structures require some definite model to describe the anisotropic creep properties of weld metal under the complex stress state. The model [80] takes into account different creep and damage material properties in longitudinal and transversal directions of a welding seam. Thus, the engineering creep theory have been applied for development of the creep-damage model based on creep potential and flow rule, e.g. creep theory proposed by Betten [34]. Experimental data [94] have been processed and used for the development of the transversally-isotropic creep model formulated with stress tensor invariants as shown below. Origins and theoretical bases of the investigated creep model are described in details in [149, 151, 152].

Stress tensor for the anisotropic material of weld metal can be decomposed as illustrated on Fig. 2.5 in the following form

$$\sigma = \sigma_{mm}\, \boldsymbol{m} \otimes \boldsymbol{m} + \sigma_p + \boldsymbol{\tau}_m \otimes \boldsymbol{m} + \boldsymbol{m} \otimes \boldsymbol{\tau}_m, \qquad (2.2.15)$$

into its corresponding projections

$$\begin{aligned}
\sigma_{mm} &= \boldsymbol{m} \cdot \sigma \cdot \boldsymbol{m}, \\
\sigma_p &= (\boldsymbol{I} - \boldsymbol{m} \otimes \boldsymbol{m}) \cdot \sigma \cdot (\boldsymbol{I} - \boldsymbol{m} \otimes \boldsymbol{m}), \\
\boldsymbol{\tau}_m &= \boldsymbol{m} \cdot \sigma \cdot (\boldsymbol{I} - \boldsymbol{m} \otimes \boldsymbol{m}).
\end{aligned} \qquad (2.2.16)$$

2.2. Anisotropic creep-damage models

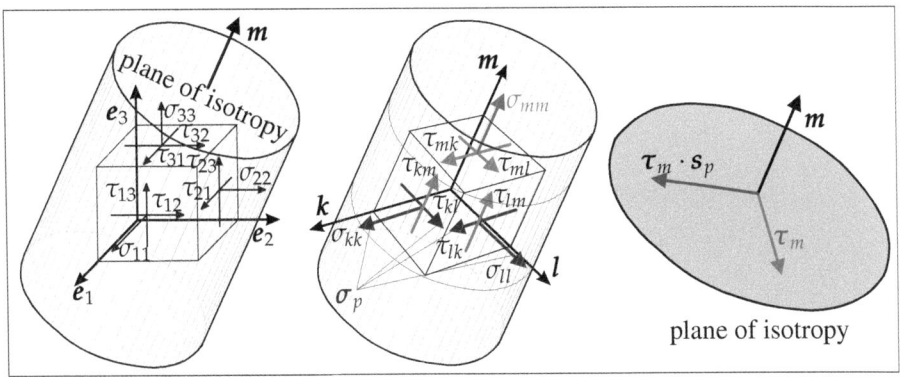

Figure 2.5: Stress state in a transversely isotropic medium and corresponding projections $\sigma_{mm}, \boldsymbol{\sigma}_p$ and $\boldsymbol{\tau}_m$, after [151, 152].

In notations (2.2.15) - (2.2.16) $\boldsymbol{m} = \boldsymbol{e}_i m_i$ is the column vector of welding direction, $\boldsymbol{I}$ is the second rank identity tensor, σ_{mm} is the normal stress acting in the plane with the unit normal $\boldsymbol{m}$, $\boldsymbol{\sigma}_p$ stands for the "plane" part of the stress tensor $\boldsymbol{\sigma}$ representing the stress state in the isotropy plane, $\boldsymbol{\tau}_m$ is the shear stress vector in the plane with the unit normal $\boldsymbol{m}$. For the orthonormal basis $\boldsymbol{k}, \boldsymbol{l}$ and $\boldsymbol{m}$ the projections are (see Fig. 2.5):

$$\boldsymbol{\tau}_m = \tau_{mk} \boldsymbol{k} + \tau_{ml} \boldsymbol{l},$$
$$\boldsymbol{\sigma}_p = \sigma_{kk} \boldsymbol{k} \otimes \boldsymbol{k} + \sigma_{ll} \boldsymbol{l} \otimes \boldsymbol{l} + \tau_{kl} (\boldsymbol{k} \otimes \boldsymbol{l} + \boldsymbol{l} \otimes \boldsymbol{k}). \qquad (2.2.17)$$

The plane part $\boldsymbol{\sigma}_p$ of the stress tensor $\boldsymbol{\sigma}$ can be further decomposed as follows

$$\boldsymbol{\sigma}_p = \boldsymbol{s}_p + \frac{1}{2} \operatorname{tr} \boldsymbol{\sigma}_p (\boldsymbol{I} - \boldsymbol{m} \otimes \boldsymbol{m}), \quad \operatorname{tr} \boldsymbol{s}_p = 0, \qquad (2.2.18)$$

where $\boldsymbol{s}_p$ is the "plane" part of the stress tensor deviator $\boldsymbol{s}$ in the plane with the unit normal $\boldsymbol{m}$.

It is possible to introduce the following set of transversely isotropic invariants, as

2. Conventional approach to creep-damage modeling

proposed in [149, 151, 152]:

$$I_{1m} = \sigma_{mm} = \boldsymbol{m} \cdot \boldsymbol{\sigma} \cdot \boldsymbol{m},$$

$$I_{2m} = \text{tr}\, \boldsymbol{\sigma}_p = \text{tr}\, \boldsymbol{\sigma} - \boldsymbol{m} \cdot \boldsymbol{\sigma} \cdot \boldsymbol{m},$$

$$I_{3m} = \tfrac{1}{2} \text{tr}\, \boldsymbol{s}_p^2 = \tfrac{1}{2} \left(\text{tr}\, \boldsymbol{\sigma}^2 - 2\, \boldsymbol{m} \cdot \boldsymbol{\sigma}^2 \cdot \boldsymbol{m} + (\boldsymbol{m} \cdot \boldsymbol{\sigma} \cdot \boldsymbol{m})^2 - \tfrac{1}{2}\left(\text{tr}\, \boldsymbol{\sigma}_p\right)^2 \right), \quad (2.2.19)$$

$$I_{4m} = \boldsymbol{\tau}_m \cdot \boldsymbol{\tau}_m = \boldsymbol{m} \cdot \boldsymbol{\sigma}^2 \cdot \boldsymbol{m} - (\boldsymbol{m} \cdot \boldsymbol{\sigma} \cdot \boldsymbol{m})^2,$$

$$I_{5m} = \boldsymbol{\tau}_m \cdot \boldsymbol{s}_p \cdot \boldsymbol{\tau}_m, \quad I_{6m} = \boldsymbol{m} \cdot (\boldsymbol{\tau}_m \cdot \boldsymbol{s}_p \times \boldsymbol{\tau}_m).$$

Taking into account the assumption $J_m \equiv I_{1m} - \tfrac{1}{2} I_{2m}$ similar to the variant of isotropic creep in [161], one can introduce the equivalent stress built on the transversely isotropic invariants of stress state as follows

$$\sigma_{eq}^2 = J_m^2 + 3\,\alpha_1\, J_{3m} + 3\,\alpha_2\, J_{4m}, \quad (2.2.20)$$

where α_1 and α_2 are material creep parameters of the transversely isotropic medium. If that parameters take the values $\alpha_1 = 1$ and $\alpha_2 = 1$ than the equivalent stress in the form (2.2.20) gets the classical form of von Mises effective stress σ_{vM} for an isotropic material.

Under the assumption that the net area reduction due to cavity formation proceeds mainly on the planes perpendicular to the direction of the maximum tensile effective stress σ_I, and the rate of this cavity formation is governed by an equivalent stress measure σ_{eq} one can formulate the equations for the equivalent damage affected stresses along the welding direction $\sigma_{eq}^{\omega_1}$ and in the plane perpendicular to the welding direction $\sigma_{eq}^{\omega_2}$. Analogously to Eq. (2.1.13) for an isotropic material [121] the equivalent damage affected stresses take the following form

$$\sigma_{eq}^{\omega_1} = \beta_1\, I_{1m} + (1 - \beta_1)\, \sigma_{eq} \quad \text{and} \quad \sigma_{eq}^{\omega_2} = \beta_2\, \frac{\sigma_I}{2} + (1 - \beta_2)\, \sigma_{eq}, \quad (2.2.21)$$

where σ_I is the first principal stress.

Anisotropic creep-damage models are generally formulated using the the concept of the effective stress tensor [126]. In the case of uni-axial stress state the effective stress $\tilde{\sigma}$ is defined by Eq. (1.4.15) as proposed in [173]

$$\tilde{\sigma} = \frac{\sigma}{1 - \omega}, \quad (2.2.22)$$

2.2. Anisotropic creep-damage models

where ω is the scalar damage parameter characterizing the damage state of a material loaded with a tensile stress σ.

For the multi-axial stress states the tertiary creep stage is described by the way of the second rank damage tensor Ω assuming the directional character of damage mechanism in the weld metal. In the case of transversely isotropic creep behaviour of the weld metal the evolutional damage equation is formulated in consideration of simple loading conditions. And the second rank damage tensor Ω is replaced with two scalar damage parameters: ω_1 (along the welding direction) and ω_2 (in the plane perpendicular to the welding direction), refer to [80].

Therefore, effective stress tensor considering two scalar damage parameters ω_1 and ω_2 can be formulated as proposed in [49, 150] in the following form

$$\tilde{\sigma} = \frac{I_{1m}}{1 - \omega_1} \boldsymbol{m} \otimes \boldsymbol{m} + \frac{\sigma_p}{1 - \omega_2} + \frac{1}{\sqrt{(1 - \omega_1)(1 - \omega_2)}} (\boldsymbol{\tau}_m \otimes \boldsymbol{m} + \boldsymbol{m} \otimes \boldsymbol{\tau}_m). \qquad (2.2.23)$$

The corresponding damage evolutional equations for the scalar damage parameters ω_1 and ω_2 with an assumption (2.2.21) are formulated as proposed in [80] in the following form:

$$\dot{\omega}_1 = \frac{b_1 \left(\langle \sigma_{eq}^{\omega_1} \rangle\right)^k}{(1 - \omega_1)^{l_1}} \quad \text{and} \quad \dot{\omega}_2 = \frac{b_2 \left(\langle \sigma_{eq}^{\omega_2} \rangle\right)^k}{(1 - \omega_2)^{l_2}}. \qquad (2.2.24)$$

Finally, the constitutive equation of steady-state creep with the consideration of transversely isotropic damage accumulation and the corresponding set of creep material parameters is obtained in the form, proposed in [149–151] as follows

$$\dot{\varepsilon}^{cr} = \frac{3}{2} a \tilde{\sigma}_{eq}^{n-1} \cdot \left[\tilde{J}_m \left(\boldsymbol{m} \otimes \boldsymbol{m} - \frac{1}{3} \boldsymbol{I} \right) + \alpha_1 \tilde{s}_p + \alpha_2 \left(\tilde{\boldsymbol{\tau}}_m \otimes \boldsymbol{m} + \boldsymbol{m} \otimes \tilde{\boldsymbol{\tau}}_m \right) \right], \qquad (2.2.25)$$

where the secondary creep material parameters $a = 1.146 \cdot 10^{-22}$ [MPa^{-n}/h] and n = 8.644 are taken from [94] as constants in Norton's law for longitudinal direction of the 9CrMoNbV weld metal at 650°C; the tertiary creep material parameters b_1 = $1.6 \cdot 10^{-20}$ [h · MPa^k], $l_1 = 11.46$, $b_2 = 4.76 \cdot 10^{-20}$ [h · MPa^k], $l_2 = 20$ and $k = 7.9$

are identified using assumptions presented in [150]; correlation material parameters between longitudinal and transverse weld metal directions identified in [80] for creep properties are $\alpha_1 = 1.23$, $\alpha_2 = 1$ and for damage properties are $\beta_1 = 0.59$, $\beta_2 = 0.23$.

It was proved that the shown above transversely isotropic creep-damage model well describes the experimental creep curves [94] along the welding direction and across the welding seam at temperature 650°C in stress range from 70 to 100 MPa. The necessary material creep parameters for welding zones with different microstructure were estimated from fitting the experimental data referenced in [94, 95]. Anisotropic model for the weld metal material and conventional isotropic model by Kachanov-Rabotnov [88] for another materials of welded structure were were applied for FE-bases creep analyses in CAE-system ANSYS to describe the creep and damage processes, refer to [80]. These creep-damage models were inserted in the FORTRAN codes of subroutines incorporated into the FE-codes of ANSYS for the purpose of long-term strength analyses. This procedure is described in [22].

2.2.2 Murakami-Ohno creep model with damage induced anisotropy

In the general case of complex stress states by constant or cyclic loading conditions the grain boundary cavitation may induce the anisotropic tertiary creep response with significant dependence on the loading orientation. The anisotropic behaviour induced by damage has been observed in creep tests under nonproportional loading [143] or in tests on predamaged specimens for different loading orientations [37]. Since the nature of damage is generally anisotropic the isotropic phenomenon of the material damage may be treated as a special case of the damage anisotropy. Conversely, the isotropic damage models can be extended to models considering the damage induced anisotropy [8, 9, 33, 56, 142]. The anisotropic creep behaviour induced by damage can be characterised by introducing a tensor-valued internal state variable. A variety of phenomenological material models have been proposed with different definitions of damage tensors and corresponding evolution equations, a review can be found for example in [191].

Till now, there is no unified phenomenological approach in modeling the damage induced anisotropy. Because of a limited number of experimental data which allows a verification of anisotropic creep-damage models it is difficult to conclude about

2.2. Anisotropic creep-damage models

the applicability of different creep damage models. In what follows let us discuss the anisotropic damage concept proposed by Murakami and Ohno [142] in order to conclude about the influence of damage induced anisotropy on the long-term predictions in high-temperature power plant components, see e.g. [13, 79]. Anisotropic damage models are generally formulated using the concept of the effective stress tensor [126]. In the uni-axial case this concept was formulated by Kachanov and Rabotnov in [100, 173] in the form of effective stress $\tilde{\sigma}$, refer to Eq. (1.4.14) in Sect. 1.4.

By postulating that the principal mechanical effect of creep-damage results in the net area reduction caused by cavity formation in materials, Murakami and Ohno [141, 142] described the damage state by means of a second rank symmetric damage tensor Ω specified by the three-dimensional cavity-area density, and developed a continuum theory of creep and creep-damage of metals and alloys. The stress tensor σ is magnified to the following effective stress tensor

$$\tilde{\sigma} = \frac{1}{2}(\sigma \cdot \Phi + \Phi \cdot \sigma) \quad \text{with} \quad \Phi = [I - \Omega]^{-1}, \tag{2.2.26}$$

where I is a second rank identity tensor and Ω is a second rank symmetric damage tensor.

The results of the metallographic observations on copper and steels show that cavities caused by creep-damage develop mainly on the grain boundaries perpendicular to the maximum tensile stress [120]. Under the assumption that the net area reduction due to cavity formation proceeds mainly on the planes perpendicular to the direction of the maximum tensile effective stress $\tilde{\sigma}_I$, and the rate of this cavity formation is governed by an equivalent stress measure one can formulate analogously to Eqs (2.1.13) - (2.1.14):

$$\sigma_{eq}^{\Omega} = \alpha \tilde{\sigma}_I + (1 - \alpha) \tilde{\sigma}_{vM}, \quad \tilde{\sigma}_{vM} = \left[\frac{3}{2} \tilde{s} \cdot \cdot \tilde{s}\right]^{\frac{1}{2}}, \tag{2.2.27}$$

$$\langle \sigma_{eq}^{\Omega} \rangle = \sigma_{eq}^{\Omega} \quad \text{for} \quad \sigma_{eq}^{\Omega} > 0, \quad \langle \sigma_{eq}^{\Omega} \rangle = 0 \quad \text{for} \quad \sigma_{eq}^{\Omega} \leq 0,$$

where $\tilde{s}$ denotes the deviator of the effective stress tensor $\tilde{\sigma}$ and $\tilde{\sigma}_I$ is the maximum positive principal value of $\tilde{\sigma}$. The evolution equation the damage tensor can be formulated as follows

$$\dot{\Omega} = B \left[\langle \sigma_{eq}^{\Omega} \rangle\right]^l [\operatorname{tr}(\Phi \cdot \mathbf{n}_I \otimes \mathbf{n}_I)]^{k-l} \mathbf{n}_I \otimes \mathbf{n}_I, \tag{2.2.28}$$

2. Conventional approach to creep-damage modeling

where $\boldsymbol{n}_I$ is the principal direction, which corresponds to the first principal stress $\tilde{\sigma}_I$, and tr is the trace operation, which denote the following mathematical operations:

$$\operatorname{tr} \boldsymbol{A} = \boldsymbol{A} \cdot \cdot \boldsymbol{I} = \boldsymbol{I} \cdot \cdot \boldsymbol{A}. \tag{2.2.29}$$

In order to describe the deformation of damaged materials depending on both the net area reduction and its three-dimensional arrangement Murakami in [142] discussed a modified stress tensor $\hat{\boldsymbol{\sigma}}$ for the constitutive equation of damaged materials. Assuming a small value of the cavity area fraction and noting the condition that $\hat{\boldsymbol{\sigma}}$ should coincide with $\boldsymbol{\sigma}$ in the undamaged state, $\hat{\boldsymbol{\sigma}}$ is assumed in the following form

$$\hat{\boldsymbol{\sigma}} = \alpha_s \boldsymbol{\sigma} + \frac{1}{2} \beta_s \left(\boldsymbol{\sigma} \cdot \boldsymbol{\Phi} + \boldsymbol{\Phi} \cdot \boldsymbol{\sigma} \right) + \frac{1}{2} \left(1 - \alpha_s - \beta_s \right) \left(\boldsymbol{\sigma} \cdot \boldsymbol{\Phi}^2 + \boldsymbol{\Phi}^2 \cdot \boldsymbol{\sigma} \right), \tag{2.2.30}$$

where α_s and β_s are material constants. Inserting $\hat{\boldsymbol{\sigma}}$ and its deviatoric part $\hat{\boldsymbol{s}}$ into the Norton's law the creep constitutive equation with respect to the damage-induced anisotropy may be specified as proposed in [13] in following form

$$\dot{\boldsymbol{\varepsilon}}^{\mathrm{cr}} = \frac{3}{2} \frac{\dot{\varepsilon}^{\mathrm{cr}}_{\mathrm{eq}}}{\hat{\sigma}_{\mathrm{vM}}} \hat{\boldsymbol{s}} \quad \text{and} \quad \dot{\varepsilon}^{\mathrm{cr}}_{\mathrm{eq}} = A \left(\hat{\sigma}_{\mathrm{vM}} \right)^n, \tag{2.2.31}$$

where the modified von Mises effective stress is

$$\hat{\sigma}_{\mathrm{vM}} = \left[\frac{3}{2} \hat{\boldsymbol{s}} \cdot \cdot \hat{\boldsymbol{s}} \right]^{\frac{1}{2}}. \tag{2.2.32}$$

Chapter 3

Non-isothermal creep-damage model for a wide stress range

Many materials exhibit a stress range dependent creep behavior. The power-law creep observed for a certain stress range changes to the viscous type creep if the stress value decreases. Recently published creep experimental data for advanced heat resistant steels indicate that the high creep exponent (in the range 5-12 for power-law behaviour) may gradually decrease to the low value of approximately 1 within the stress range relevant for engineering structures. The first aim of work presented in Sect. 3.1 is to confirm the assumption, that the creep behavior is stress range dependent demonstrating the power-law to viscous transition with a decreasing stress. The assumption is based on the available creep and stress relaxation experiments for the several 9-12%Cr heat-resistant steels. An extended constitutive model for the minimum creep rate is introduced to consider both the linear and the power law creep ranges depending upon the stress level. To take into account the primary creep behavior a strain hardening function is utilized. The data for the minimum creep rate is well-defined only for moderate and high stress levels. To reconstruct creep rates for the low stress range the data of the stress relaxation test are applied, as presented in Sect. 3.2 The results show a gradual decrease of the creep exponent with the decrease stress level. Furthermore, they illustrate that the proposed constitutive model with strain hardening function well describes the creep rates for a wide stress range including the loading values relevant to engineering applications. The proposed creep constitutive model is extended with the temperature dependence using

3. Non-isothermal creep-damage model for a wide stress range

the Arrhenius-type functions.

To characterize the tertiary creep behaviour and fracture the proposed constitutive equation is generalized by introduction of damage internal state variables and appropriate evolution equations. The description of long-term strength behaviour for advanced heat-resistant steels is based on the assumption of ductile to brittle damage character transition with a decrease of stress. The second aim of work presented in Sect. 3.3 is to formulate the long-term strength equation based on ductile to brittle damage transition and to confirm it with the available creep-rupture experiments for the 9-12%Cr advanced heat-resistant steels. Therefore, two damage parameters were introduced with ductile and brittle damage accumulation characters. They are based on the same long-term strength equation describing time-to-rupture, but lead to the different types of fracture. Ductile fracture with progressive deformation and necking occurs at high stress levels and is accompanied by power-low creep deformations. However, brittle fracture caused by thermal exposure and material microstructure degradation occurs at low stress levels and it is dominantly accompanied by the linear creep deformations. And two corresponding ductile and brittle damage evolution equations based on Kachanov-Rabotnov concept are formulated for the coupling with creep constitutive equation. The proposed damage evolution equations are extended with the temperature dependence using the Arrhenius-type functions. All the necessary tertiary creep material parameters identified fitting the available from literature creep-rupture experimental data for the 9-12%Cr advanced heat-resistant steels at several temperatures.

Finally in Sect. 3.4 the unified multi-axial form of non-isothermal creep-damage model for a wide stress is proposed. The available creep-rupture experiments for 9-12%Cr heat-resistant steels [155] suggest that the brittle damage evolution is primarily controlled by the maximum tensile stress at low stress levels. And within the range of high stresses the primarily ductile creep damage is governed by the von Mises effective stress and leads to the necking of uniaxial specimen. To analyse the failure mechanisms under multi-axial stress states the isochronous rupture loci (plots of stress states leading to the same time to fracture) and time-to-rupture surface are presented. They show that the proposed failure criterion include both the maximum tensile stress and the von Mises effective stress, as in the Sdobyrev [182] and Leckie-Hayhurst criteria [85]. But the measures of influence of the both stress parameters are dependent on the level of stress showing ductile to brittle failure transition.

3.1 Non-isothermal creep constitutive modeling

Many components of power generation equipment and chemical refineries are subjected to high temperature environments and complex loading over a long time. For such conditions the structural behavior is governed by various time-dependent processes including creep deformation, stress relaxation, stress redistribution as well as damage evolution in the form of microcracks, microvoids, and other defects. The aim of "creep mechanics" is the development of methods to predict time-dependent changes of stress and strain states in engineering structures up to the critical stage of creep rupture, see e.g. [34, 152]. To this end various constitutive models which reflect time-dependent creep deformations and processes accompanying creep like hardening/recovery and damage have been recently developed. One feature of the creep constitutive modeling is the response function of the applied stress which is usually calibrated against the experimental data for the minimum (secondary) creep rate. An example is the Norton-Bailey law (power law) which is often applied because of easier identification of material constants, mathematical convenience in solving structural mechanics problems and possibility to analyze extreme cases of linear creep or perfect plasticity by setting the creep exponent to unity or to infinity, respectively. Therefore, the majority of available solutions within the creep structural mechanics are based on the power law creep assumption, e.g. [34, 40, 93, 166].

On the other hand, it is known from the materials science that the "power law creep mechanism" operates only for a specific stress range and may change to the linear, e.g. diffusion type mechanism with a decrease of the stress level [70]. As the recently published experimental data show [108, 110, 117, 178, 179], advanced heat resistant steels exhibit the transition from the power law to the linear creep at the stress levels relevant for engineering applications. To establish creep behavior for low and moderate stress levels special experimental techniques were employed. Furthermore, experimental analysis of creep under low stress values requires expensive long-time tests. Although the results presented in [108, 110, 117, 178, 179] indicate that the power law may essentially underestimate the creep rate, experimental data for many other materials is not available, and the power law stress function is usually preferred.

Generally, the ranges of "low" and "moderate" stresses are specific for many engineering structures under in-service loading conditions, e.g. [152]. The reference

3. Non-isothermal creep-damage model for a wide stress range

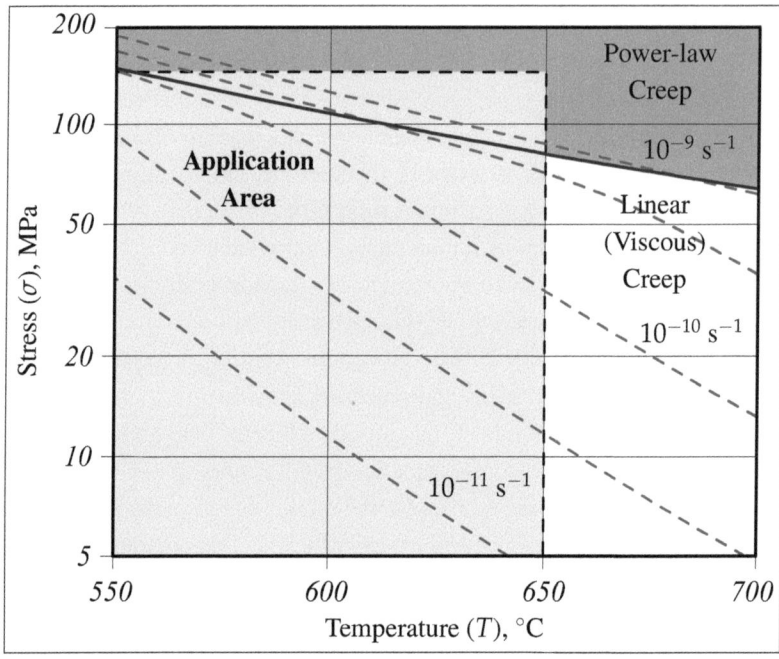

Figure 3.1: Creep deformation mechanism map of 9Cr-1Mo-V-Nb (ASTM P91) steel, after [107, 109].

stress state in a structure may significantly change during the creep process. Stresses may slowly relax down during the service time, and so the application of the power law might be questionable.

The attempt to extend the constitutive model to predict creep strains and the damage accumulation in the both ranges of "power-low creep" and "diffusional flow" is cited in [12, 153] and is presented below. The proposed phenomenological approach to creep modeling is based on deformation mechanism maps for metals (see Fig. 1.5) and available experimental creep data for heat-resistant low-alloy and high-alloy steels (see Tables 1.1 and 1.2 and Fig. 1.8).

One example of typical heat-resistant high-alloy steel is the steel 9Cr-1Mo-V-Nb (ASTM P91) or simply steel type P91. For this steel a lot of creep experimental data has been recently published, e.g. refer to [71, 72, 105–110, 165, 187–189]. In addition, it is widely spread for high temperature power plant applications, such as steam

3.1. Non-isothermal creep constitutive modeling

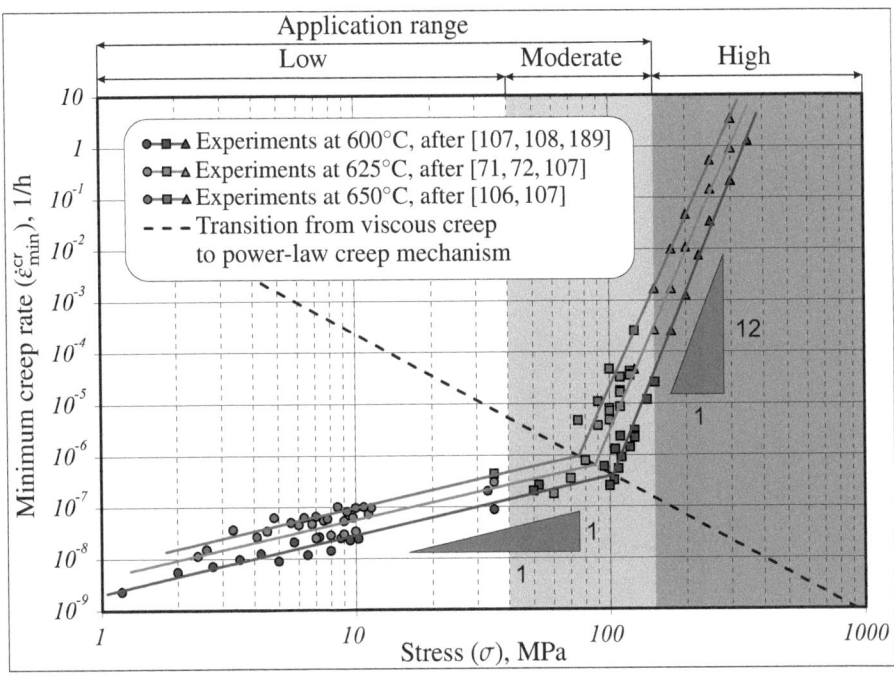

Figure 3.2: Dependence of minimum creep strain rate $\dot{\varepsilon}^{cr}_{min}$ on stress σ for 9Cr-1Mo-V-Nb (ASTM P91) heat-resistant steel derived from the experiments after [71, 72, 106–108, 189].

turbine components and high pressure piping. Creep deformation mechanism map of P91 steel (see Fig. 3.1) shows that it is necessary to take into account both creep mechanisms during the creep modeling: power-law creep, which includes generally high stress range, and linear or viscous creep, which includes generally moderate and low stress ranges, refer to [107, 109].

Within the phenomenological approach to creep modeling one usually starts with the constitutive equation for the minimum (secondary) creep rate $\dot{\varepsilon}^{cr}_{min}$. Figure 3.2 illustrates all collected experimental data for 9Cr-1Mo-V-Nb (ASTM P91) heat-resistant steel, presenting the dependence of the minimum creep rate on the applied stress for different temperatures, where the ranges of "low", "moderate" and "high" stress values are explained. Here the transition stress σ_0 reflects the transition from linear (viscous) creep mechanism to power-law creep mechanism.

3. Non-isothermal creep-damage model for a wide stress range

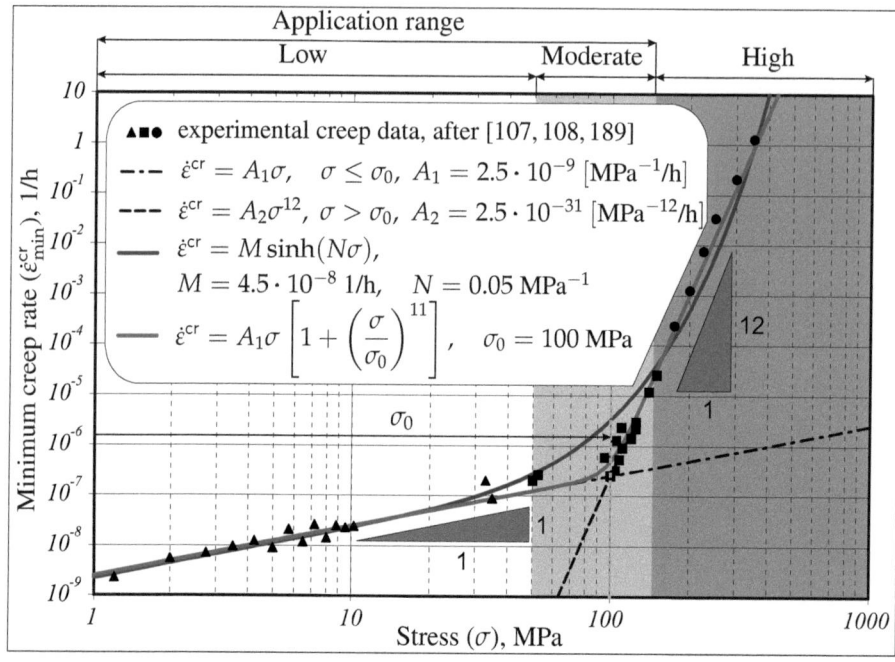

Figure 3.3: Fitting of creep experimental data [107, 108, 189] of minimum creep strain rate $\dot{\varepsilon}^{cr}$ by different constitutive equations for 9Cr-1Mo-V-Nb (ASTM P91) heat-resistant steel at 600°C.

We have decided to concentrate our investigations on temperature 600°C as the most spread in-service temperature among power plant components made of 9Cr-1Mo-V-Nb (ASTM P91) heat resistant steel, see Fig. 3.3. Within the "low" stress range the creep rate is nearly linear function of stress, which corresponds to viscous creep character. The "moderate" stress range is characterized by the transition from linear to power law dependence. Within the region of "high" stresses the value of the creep exponent is usually in the range between 4 and 12 depending on the material, type of alloying and processing conditions. The experimental data for 9Cr-1Mo-V-Nb (ASTM P91) heat resistant steel show that the high stress range creep exponent takes the value 12.

At first, the hyperbolic sine (*Sinh*) stress response function was tried as a basis of the constitutive model suitable for complete stress range, see Fig. 3.3. It fits satis-

3.1. Non-isothermal creep constitutive modeling

factory experimental data for the "low" and "high" stress ranges, but not suitable for "moderate" stress range. Therefore, the so-called double power-law stress response function proposed in [107] has been selected as a basis. Relying on the assumptions of stress-dependent transition of creep-deformation mechanism from [107], Sect. 1.3 of the thesis and Norton-Bailey [27, 157] equation (1.3.9) the following creep constitutive equation for a fixed temperature and wide stress range is proposed

$$\dot{\varepsilon}^{cr} = \dot{\varepsilon}^{cr}_v + \dot{\varepsilon}^{cr}_{pl} = A_1 \sigma + A_2 \sigma^n, \qquad (3.1.1)$$

where $\dot{\varepsilon}^{cr}$ is the steady-state creep rate, $\dot{\varepsilon}^{cr}_v$ is the rate of viscous mechanism, $\dot{\varepsilon}^{cr}_{pl}$ is the rate of power-law mechanism and A_1, A_2 and the stress exponent n are secondary creep material parameters. Otherwise, the proposed isothermal creep constitutive model presented by Eq. (3.1.1) can be transformed to the normalized form, which is more suitable for the creep material parameters identification (for the detailed description refer to [12, 17, 78, 153]):

$$\dot{\varepsilon}^{cr} = A \sigma \left[1 + \left(\frac{\sigma}{\sigma_0} \right)^{n-1} \right], \qquad (3.1.2)$$

where σ_0 is the transition stress, presenting the transition from power-law creep behaviour to linear. The required material parameters of the 9Cr-1Mo-V-Nb (ASTM P91) heat-resistant steel for the creep constitutive model (3.1.2) are the following: $A = A_1 = 2.5 \cdot 10^{-9}$ [MPa^{-1}/h], $n = 12$, $\sigma_0 = 100$ MPa. They are easily defined from separate fitting of experimental data for "low" and "high" stress ranges. The proposed creep constitutive equation (3.1.2) reflects the transition from viscous to power-law creep mechanism and provides a good fitting of available experimental data [107, 108, 189] for the complete stress range, as illustrated on Fig. 3.3.

And the connection between the equivalent forms of creep constitutive equation presented by Eqs (3.1.1) and (3.1.2) is following

$$A_2 = \frac{A}{(\sigma_0)^{n-1}} \quad \Longleftrightarrow \quad \sigma_0 = \left(\frac{A}{A_2} \right)^{\frac{1}{n-1}}. \qquad (3.1.3)$$

The creep constitutive equation (3.1.1) can be extended with temperature dependence assuming temperature-dependent secondary creep material parameters $A_1(T)$ and $A_2(T)$ in the following form

$$\dot{\varepsilon}^{cr} = \dot{\varepsilon}^{cr}_v(T) + \dot{\varepsilon}^{cr}_{pl}(T) = A_1(T) \sigma + A_2(T) \sigma^n, \qquad (3.1.4)$$

3. Non-isothermal creep-damage model for a wide stress range

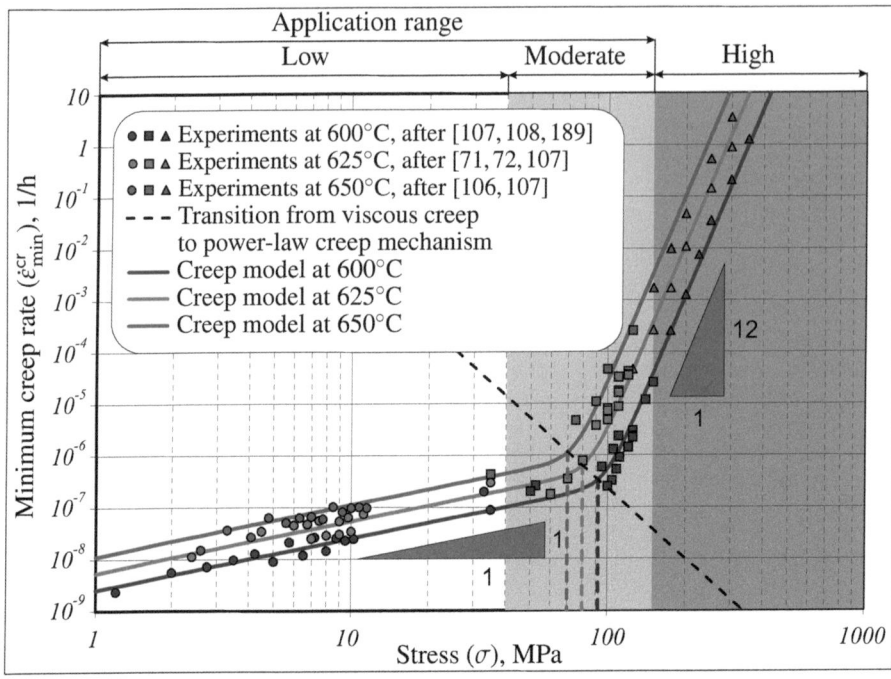

Figure 3.4: Comparison of non-isothermal creep constitutive model (3.1.6) with experiments [71, 72, 106–108, 189] for the 9Cr-1Mo-V-Nb (ASTM P91) steel at 600°C, 625°C and 650°C.

where the temperature dependence is introduced by the Arrhenius-type functions (1.3.8) similarly to Eq. (1.3.11)

$$A_1(T) = A_{01} \exp\left(\frac{-Q_1}{RT}\right) \quad \text{and} \quad A_2(T) = A_{02} \exp\left(\frac{-Q_2}{RT}\right), \quad (3.1.5)$$

where the constants A_{01}, A_{02} and creep activation energies Q_1, Q_2 present the secondary creep material parameters to be identified, and the universal gas constant $R = 8.31$ [J·K^{-1}·mol^{-1}].

The normalized form of the creep constitutive equation (3.1.2) can be also extended with temperature dependence assuming temperature-dependent secondary

3.1. Non-isothermal creep constitutive modeling

creep material parameter $A_1(T)$ and transition stress $\sigma_0(T)$ in the following form

$$\dot{\varepsilon}^{cr} = A_1(T)\,\sigma\left[1 + \left(\frac{\sigma}{\sigma_0(T)}\right)^{n-1}\right], \qquad (3.1.6)$$

where the connection between the equivalent forms of creep constitutive model expressed by Eqs (3.1.4) and (3.1.6) gives the following form of temperature-dependent transition stress:

$$A_2(T) = \frac{A_1(T)}{[\sigma_0(T)]^{n-1}} \iff \sigma_0(T) = \left[\frac{A_1(T)}{A_2(T)}\right]^{\frac{1}{n-1}}. \qquad (3.1.7)$$

Therefore, the transition stress σ_0 can be presented in simple temperature-dependent form using Arrhenius-type function (1.3.8) as follows

$$\sigma_0(T) = A_\sigma \exp\left(\frac{-Q_\sigma}{RT}\right), \qquad (3.1.8)$$

where A_σ is the secondary creep material parameter and Q_σ is the activation energy of the transition to be identified. Creep parameters for the temperature-dependent transition stress $\sigma_0(T)$ from Eq. (3.1.8) are connected with creep material parameters for linear regime (A_{01}, Q_1) and power-law regime (A_{02}, n, Q_2) from Eqs (3.1.4) and (3.1.5) in the following way

$$A_\sigma = \left(\frac{A_{01}}{A_{02}}\right)^{\frac{1}{n-1}} \quad \text{and} \quad Q_\sigma = \frac{Q_1 - Q_2}{n-1}. \qquad (3.1.9)$$

Finally, the experimental data [71, 72, 106–108, 189] presenting minimum creep rate $\dot{\varepsilon}^{cr}_{min}$ vs. stress σ for 9Cr-1Mo-V-Nb (ASTM P91) heat-resistant steel at temperatures 600°C, 625°C and 650°C is fitted with the creep constitutive equation (3.1.6). And as a result the following secondary creep material parameters applicable to the wide stress and temperature ranges are identified: $A_{01} = 2300$ [MPa^{-1}/h], $Q_1 = 200000$ [J·mol^{-1}], $A_\sigma = 0.658$ MPa, $Q_\sigma = 36364$ [J·mol^{-1}] and $n = 12$.

The comparison of non-isothermal creep constitutive model (3.1.6) with experimental data after [71, 72, 106–108, 189] for 9Cr-1Mo-V-Nb (ASTM P91) heat-resistant steel at temperatures 600°C, 625°C and 650°C shows a good correlation, as illustrated on Fig. 3.4.

3. Non-isothermal creep-damage model for a wide stress range

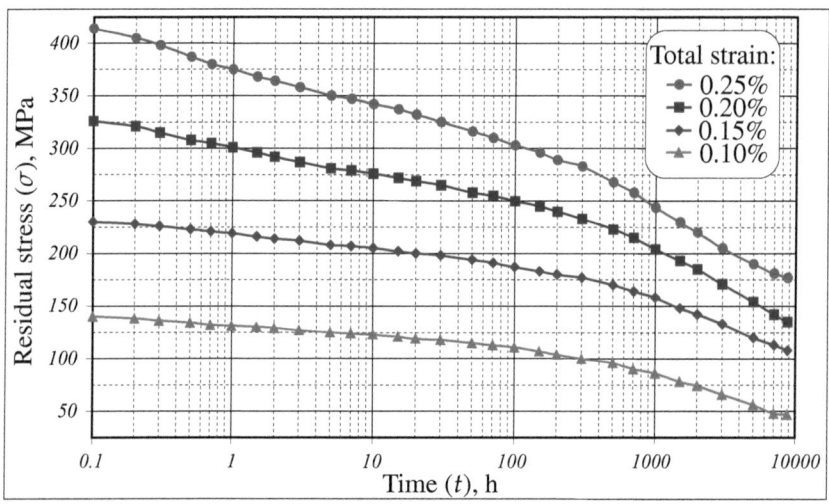

Figure 3.5: Stress relaxation experimental data for the 12Cr-1Mo-1W-0.25V steel at 500°C, after [163].

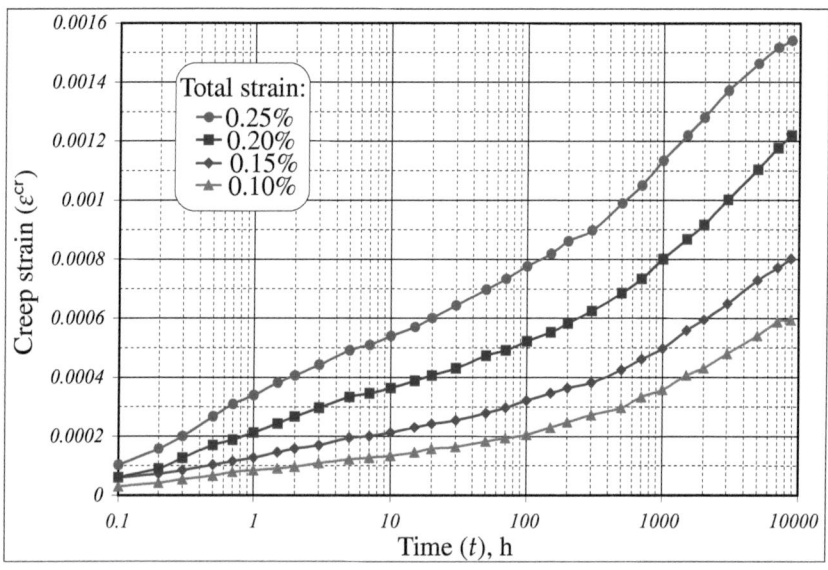

Figure 3.6: Creep strains calculated from the relaxation data for the 12Cr-1Mo-1W-0.25V steel bolting material at temperature 500°C, after [163].

3.2. Stress relaxation problem and primary creep modeling

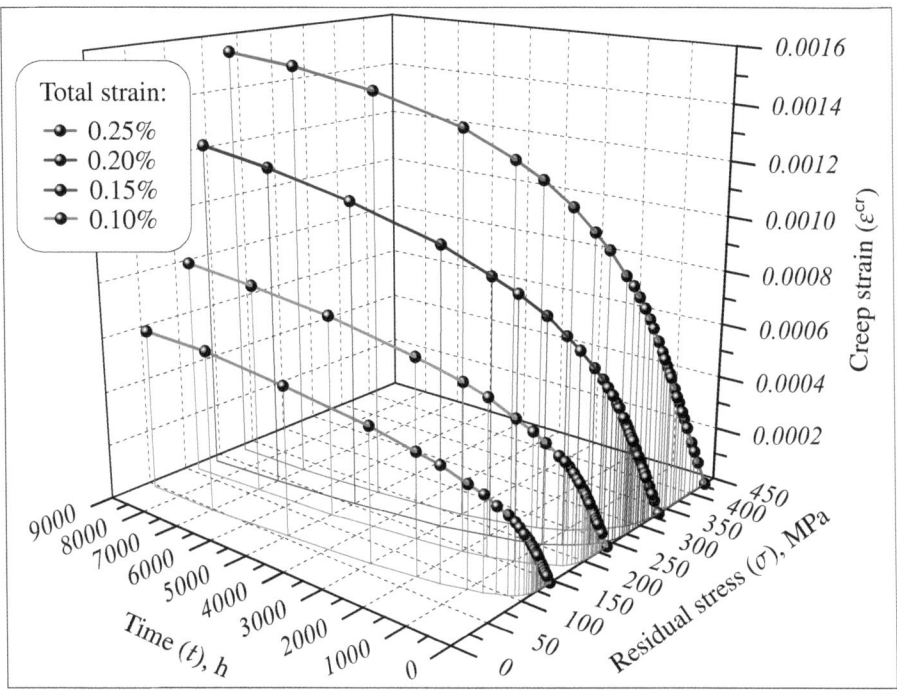

Figure 3.7: 3D trajectory plot of relaxation experimental data for 12Cr-1Mo-1W-0.25V steel bolting material at temperature 500°C, after [163].

3.2 Stress relaxation problem and primary creep modeling

In structural analysis applications it is often desirable to consider stress redistributions from the beginning of the creep process up to the creep with constant rate. Let us note, that in a statically undetermined structure stress redistributions take place even if primary creep is ignored. The creep modeling with consideration of primary creep are important in many applications of structural analysis when loading is in the ranges of moderate and low stresses, which are relevant for engineering structures. The examples are stress relaxation and creep analysis for a wide stress range [17, 66, 78], multi-axial creep of thin-walled tubes under combined action of tension (compression) force and torque [152], analysis of the primary creep

3. Non-isothermal creep-damage model for a wide stress range

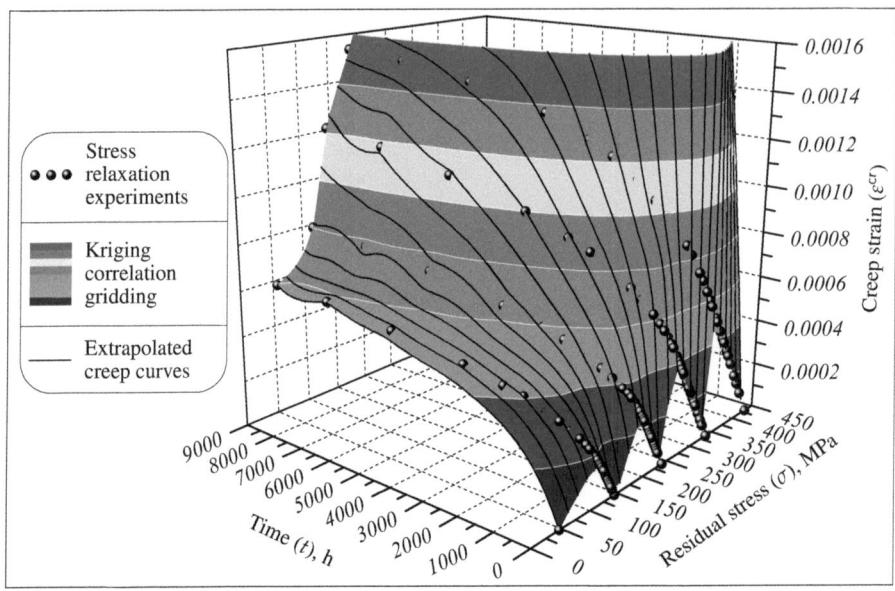

Figure 3.8: 3D interpolated surface fitting the relaxation experiments [163] for the 12Cr-1Mo-1W-0.25V steel at 500°C.

behaviour of thin-walled shells subjected to internal pressure [35], the creep buckling of cylindrical shells subjected to internal pressure and axial compression [36].

The aim of this Section is to confirm the stress range dependence of creep behavior based on the experimental data of stress relaxation. An extended constitutive model for the minimum creep rate is introduced to consider both the linear and the power law creep ranges. To take into account the primary creep behavior a strain hardening function is utilized. The material constants are identified basing on the published experimental data of creep and stress relaxation for a 9-12%Cr advanced heat-resistant steels [66, 163, 164]. The data for the minimum creep rate is well-defined only for moderate and high stress levels. To reconstruct creep rates for the low stress range the data of the stress relaxation test are applied. The results show a gradual decrease of the creep exponent with the decrease stress level. Furthermore, they illustrate that the proposed constitutive model well describes the creep rates for a wide stress range.

3.2. Stress relaxation problem and primary creep modeling

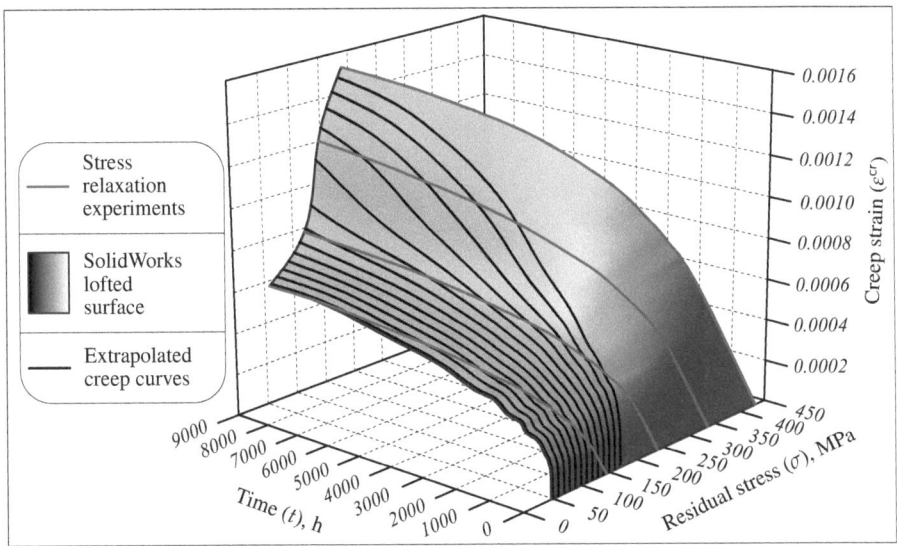

Figure 3.9: 3D lofted surface through the trajectories of relaxation experimental data for 12Cr-1Mo-1W-0.25V steel bolting material at temperature 500°C, after [163].

3.2.1 Stress relaxation problem

Primary and secondary creep behaviour of modern heat-resistant steels can be analysed using the available stress relaxation curves obtained from uniaxial experiments under the high temperature conditions, see e.g. [66], as shown below. Stress relaxation and creep experimental data for the 12Cr-1Mo-1W-0.25V steel bolting material at 500°C presented in [163, 164] has been selected for the formulation of the constitutive creep model for a wide stress range. Figure 3.5 illustrates the typical stress relaxation curves of the 12Cr-1Mo-1W-0.25V steel at 500°C obtained during the experiments [163]. The uniaxial constitutive model presented below is valid for primary and secondary creep stages and includes the creep material parameters identified from the available stress relaxation and creep experimental data.

Experimental relaxation curves [163] can be transformed into creep strain vs. time dependence (see Fig. 3.6) as follows:

$$\varepsilon^{\mathrm{cr}}(t) = \frac{\sigma_{\mathrm{ini}} - \sigma(t)}{E}, \quad (3.2.10)$$

3. Non-isothermal creep-damage model for a wide stress range

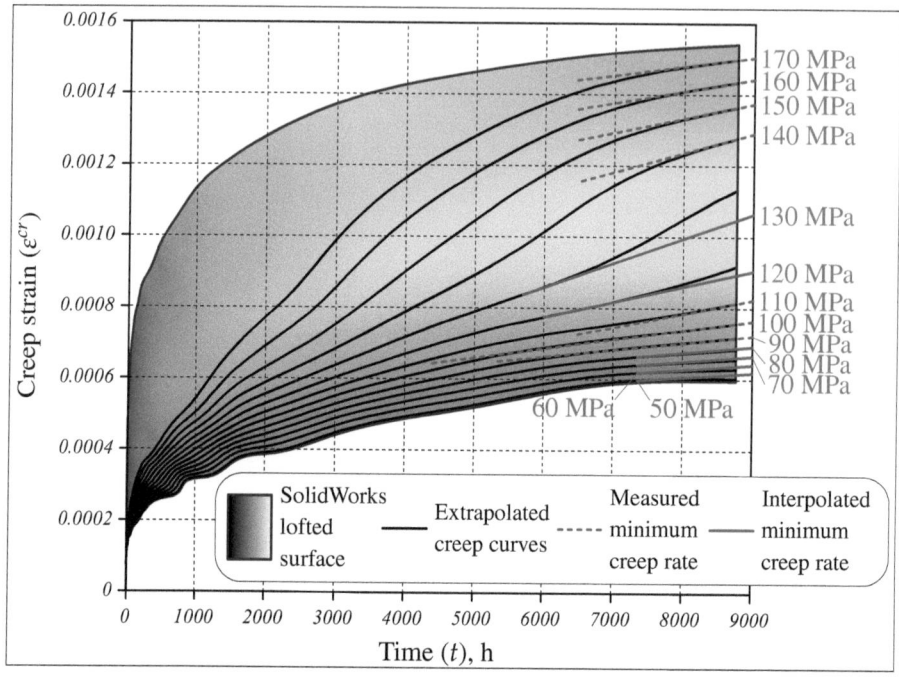

Figure 3.10: Creep curves derived from 3D lofted surface created in SolidWorks as cross-sections perpendicular to residual stress axis.

where $\varepsilon^{cr}(t)$ denotes the creep strain, σ_{ini} is the initial stress in the test, $\sigma(t)$ is the residual stress (the experimental values of σ_{ini} and $\sigma(t)$ are given in [163]), $E = 164.8$ GPa is the Young's modulus of the 12Cr-1Mo-1W-0.25V steel at 500°C from [194].

The experimental dependencies presenting residual stress vs. time and creep strain vs. time can be combined into a general 3D plot with the time t, the residual stress σ and the creep strain ε^{cr} as orthogonal axes containing relaxation trajectories corresponding to the defined values of the constant total strain $\varepsilon^{tot} = 0.25\%, 0.20\%, 0.15\%, 0.10\%$, as illustrated on Fig. 3.7. These trajectories are formed by the interpolation of experimental points with $\varepsilon^{cr}(t)$, $\sigma(t)$ and t as coordinates into solid lines in the CAD-software SolidWorks.

The relaxation experimental data for 12Cr-1Mo-1W-0.25V steel bolting material at 500°C in the form of 3D trajectories presenting creep strain ε^{cr} vs. residual stress

3.2. Stress relaxation problem and primary creep modeling

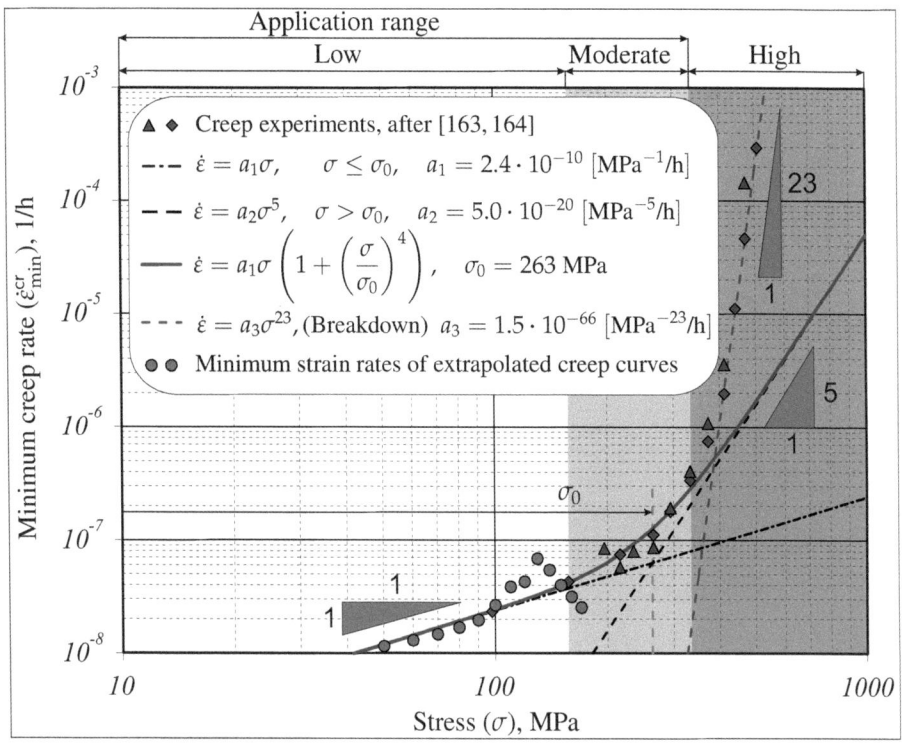

Figure 3.11: Minimum creep rate vs. stress for 12Cr-1Mo-1W-0.25V heat-resistant steel at 500°C.

σ and time t is fitted with the 3D surface by the way of kriging correlation gridding in the mathematical software OriginPro, as illustrated on Fig. 3.8.

But the 3D surface obtained by the kriging correlation gridding does not fit precisely the relaxation experimental data for 12Cr-1Mo-1W-0.25V steel bolting material at 500°C after [163]. Therefore, the experimental data in the form of the stress relaxation trajectories is fitted with the 3D lofted surface formed by the loft operation through these trajectories in the CAD-software SolidWorks, as illustrated on Fig. 3.9. Moreover, the creep curves illustrated on Fig. 3.10, which are derived from the 3D lofted surface as cross-sections perpendicular to the residual stress axis σ, have both primary and secondary stages and demonstrate the necessity to apply a combined primary and secondary uniaxial creep model to describe them.

3. Non-isothermal creep-damage model for a wide stress range

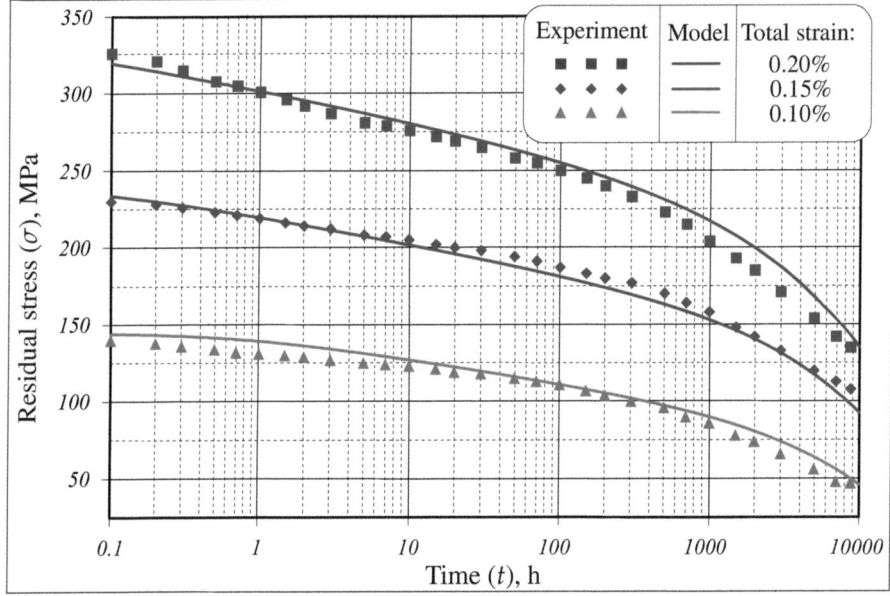

Figure 3.12: Comparison of the numerical solution for the uniaxial stress relaxation problem by the creep model (3.2.14) with experimental data [163] for the 12Cr-1Mo-1W-0.25V steel at 500°C.

To model the secondary creep behaviour for the wide stress range including both the low and the moderate stress values the following double-power-law constitutive equation (3.1.2) for the minimum creep rate $\dot{\varepsilon}^{cr}_{min}$ (for the detailed description refer to Sect. 3.1) was applied:

$$\dot{\varepsilon}^{cr} = a_1 (\sigma H) + a_2 (\sigma H)^n = A \sigma H \left[1 + \left(\frac{\sigma H}{\sigma_0} \right)^{n-1} \right], \quad (3.2.11)$$

including the strain hardening function $H(\varepsilon^{cr})$ to describe the primary creep stage, which was proposed in [114]

$$H(\varepsilon^{cr}) = 1 + \alpha\, e^{-\beta\, \varepsilon^{cr}}. \quad (3.2.12)$$

The steady-state creep material parameters of the 12Cr-1Mo-1W-0.25V steel bolting material at 500°C for the creep constitutive equation (3.2.11) were iden-

3.2. Stress relaxation problem and primary creep modeling

tified manually to fit the experimental values of minimum creep strain rate $\dot{\varepsilon}^{cr}_{min}$ from [163, 164], as illustrated on Fig. 3.11. In Eq. (3.2.11) the secondary material creep parameter $a_1 = A = 2.4 \cdot 10^{-10}$ [MPa^{-1}/h] corresponds to linear creep, the parameters $a_2 = 5.0 \cdot 10^{-20}$ [MPa^{-5}/h] and $n = 5$ correspond to power-law creep, and $\sigma_0 = 263$ MPa denotes the transition stress from viscous to power-law creep. The strain hardening function (3.2.12) contains α and β as the appropriate primary creep material parameters of the 12Cr-1Mo-1W-0.25V steel bolting material at 500°C, which are identified manually by fitting the experimental stress relaxation curves from [163].

For an accurate description of the stress relaxation the constitutive model presented by Eqs (3.2.11) - (3.2.12) which reflects both the hardening and the steady-state creep is required. The solution of the stress relaxation problem is based on the total strain ε^{tot} remaining constant and the conversion of the elastic strain ε^{el} by the stress reduction into the creep strain ε^{cr} formulated similarly to [66]:

$$-\left(\frac{1}{E}\right)\frac{d\sigma}{dt} = \frac{d\varepsilon^{cr}}{dt}, \qquad (3.2.13)$$

or taking into account Eqs (3.2.11) - (3.2.12) the relaxation problem is formulated as follows

$$-\left(\frac{1}{E}\right)\frac{d\sigma}{dt} = A\,\sigma\left(1+\alpha\,e^{-\beta\,\varepsilon^{cr}}\right)\left[1+\left(\frac{\sigma\,(1+\alpha\,e^{-\beta\,\varepsilon^{cr}})}{\sigma_0}\right)^{n-1}\right], \qquad (3.2.14)$$

where the creep strain ε^{cr} is replaced with $(\sigma_{ini} - \sigma)/E$ accordingly to Eq. (3.2.10).

The differential equation (3.2.14) is solved numerically using the MathCAD software for several values of initial conditions, i.e. initial stress values $\sigma_{ini} = 145$ MPa, 240 MPa, 336 MPa corresponding to several values of the total strain $\varepsilon^{tot} = 0.10\%$, 0.15%, 0.20% in the relaxation experiments [163], as illustrated on Fig. 3.12. During the solution of the relaxation problem (3.2.14) the values of the primary creep material parameters $\alpha = 6$ and $\beta = 4500$ from the hardening function (3.2.12) were identified for the 12Cr-1Mo-1W-0.25V steel at 500°C by fitting the experimental relaxation curves from [163]. Figure 3.12 illustrates a good agreement of the numerical solution for the stress relaxation problem (3.2.14) with the experimental data after [163] and confirms the applicability of the proposed constitutive model (3.2.11) - (3.2.12) to the description of the creep behaviour for advanced heat resistant steels.

3. Non-isothermal creep-damage model for a wide stress range

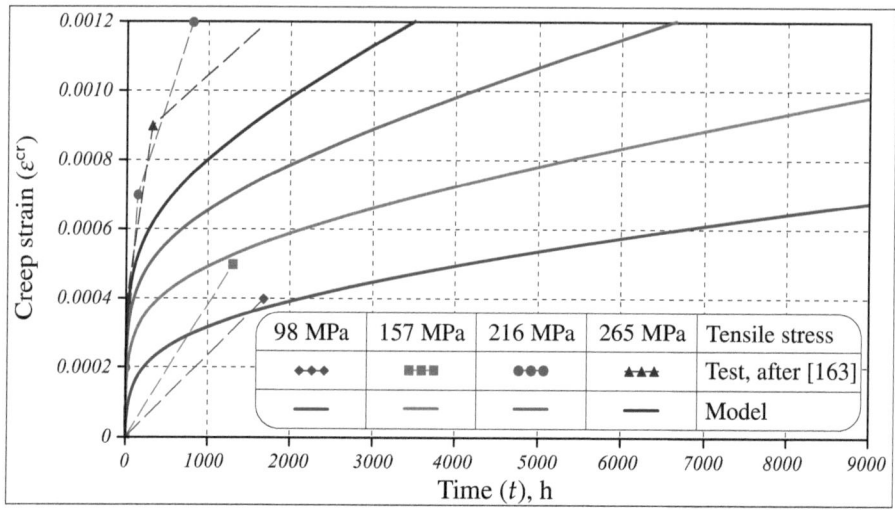

Figure 3.13: Comparison of creep curves obtained by creep constitutive model including strain hardening with experimental creep curves from [163] for the 12Cr-1Mo-1W-0.25V steel at 500°C.

3.2.2 Primary creep strain

Since there is no stress relaxation experimental data for the 9Cr-1Mo-V-Nb (ASTM P91) heat-resistant steel found in literature, then it is necessary to rely on the available creep experiments during the primary creep modeling. The primary and secondary creep stages of experimental creep curves [165, 188] for the 9Cr-1Mo-V-Nb (ASTM P91) steel at 600°C can be modeled using the creep constitutive equation (3.2.11) with previously defined in Sect. 3.1 secondary creep parameters for Eq. (3.1.2) as $A = 2.5 \cdot 10^{-9}$ [MPa^{-1}/h], $n = 12$ and $\sigma_0 = 100$ MPa. For the purpose to identify the primary creep material parameters α and β in the strain hardening function (3.2.12) included in Eq. (3.2.11), the set of experimental creep curves was selected from [165, 188]. The selected creep curves with apparent primary creep stages corresponds to several constant stress values $\sigma = 120$ MPa, 125 MPa, 150 MPa, 200 MPa. Then the set of model creep curves corresponding to the same constant stress values was created using numerical integration of constitutive equation (3.2.11) in MathCAD software. The primary creep stages of model creep curves

3.2. Stress relaxation problem and primary creep modeling

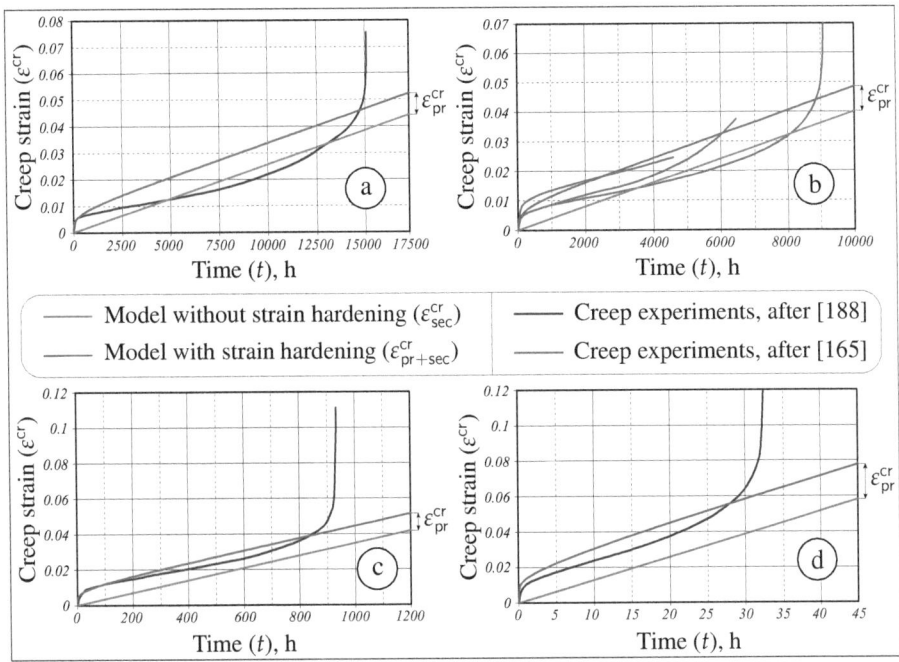

Figure 3.14: Primary creep stage fitting of the experimental creep curves after [165, 188] and definition of the primary creep strain values ε_{pr}^{cr} for the 9Cr-1Mo-V-Nb (ASTM P91) steel at 600°C for the following values of tensile stress σ: a) 120 MPa, b) 125 MPa, c) 150 MPa, d) 200 MPa.

were fitted to the correspondent experimental curves using serial iterations, as illustrated on Fig. 3.14. And in the result the values of primary creep material parameters have been found as $\alpha = 0.5$ and $\beta = 300$.

The next step was the creation of the set of model creep curves without primary creep stage using constitutive equation (3.1.2) for the same constant stress values as before, see Fig. 3.14. Consequently, the creep strain accumulated during the primary stage ε_{pr}^{cr} is calculated as the positive difference between the creep strain $\varepsilon_{pr+sec}^{cr}$ accumulated with strain hardening effect by Eq. (3.2.11) and the creep strain ε_{sec}^{cr} accumulated without strain hardening effect by Eq. (3.1.2) in the following form:

$$\varepsilon_{pr}^{cr} = \varepsilon_{pr+sec}^{cr}(\sigma) - \varepsilon_{sec}^{cr}(\sigma). \quad (3.2.15)$$

3. Non-isothermal creep-damage model for a wide stress range

It is necessary to notice, that the mathematical operation (3.2.15) was performed numerically in MathCAD software for the whole stress range at 600°C. The creep strain accumulated during the primary stage ε_{pr}^{cr} was calculated for each value of stress σ on the moments of time, when the influence of the strain hardening function (3.2.12) on the creep curve was negligible, as illustrated on Fig. 3.14.

Finally, the primary creep strain ε_{pr}^{cr} vs. stress σ dependence, illustrated on Fig. 3.15, can be fitted using the Boltzmann function producing a sigmoidal curve

$$\varepsilon_{pr}^{cr}(\sigma) = \varepsilon_{pr\,max}^{cr} + \frac{\varepsilon_{pr\,min}^{cr} - \varepsilon_{pr\,max}^{cr}}{1 + \exp\left(\frac{\sigma - \bar{\sigma}}{C}\right)}, \quad (3.2.16)$$

where the defined fitting parameters are following: the transition constant $C = 9.5$, transition stress $\bar{\sigma} = 87$ MPa corresponding to the mean value of primary creep strain $\bar{\varepsilon}_{pr}^{cr} = 4.56 \cdot 10^{-3}$, primary creep strain value in the low stress range $\varepsilon_{pr\,min}^{cr} = 1.35 \cdot 10^{-3}$, primary creep strain value in the high stress range $\varepsilon_{pr\,max}^{cr} = 7.76 \cdot 10^{-3}$.

The primary creep strain function (3.2.16) will be taken into account in Sect. 3.3 during the fitting of the experimental rupture creep strain ε^* vs. stress σ dependence by the Eq. (3.3.28) for the purpose of the tertiary creep material parameters identification.

3.3 Non-isothermal long-term strength and tertiary creep modeling

One of the approaches to long-term strength estimation of advanced heat-resistant steels is based on the time-to-failure concept in the form proposed in [25, 122, 136, 197]. Since creep causes creep fracture, the time-to-failure t^* is described by a constitutive equation which looks very like that for creep itself, see e.g. Eq. (1.3.11)

$$t^* = B_f\, \sigma^{-m} \exp\left(\frac{Q_f}{RT}\right), \quad (3.3.17)$$

where B_f, m and Q_f are the creep-failure constants, determined in the same way as those for creep. The behavior of t^* with respect to σ and T is similar to that of $\dot{\varepsilon}^{cr}$, with the differences being that the signs are reversed for the stress exponent m and the activation energy Q_f, because t^* is a time whereas $\dot{\varepsilon}^{cr}$ is a rate, refer to [25, 197].

3.3. Non-isothermal long-term strength and tertiary creep modeling

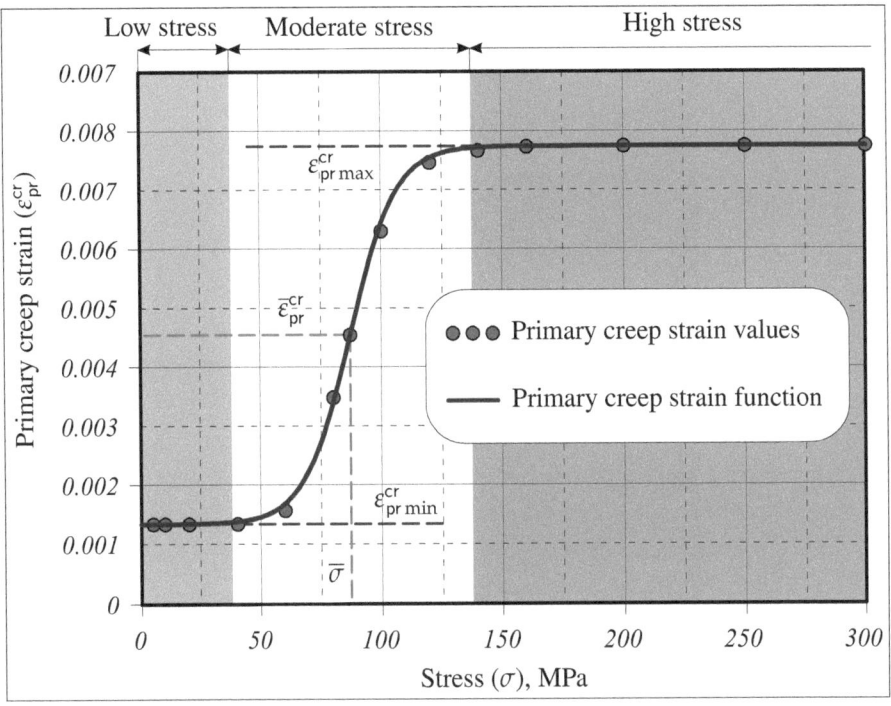

Figure 3.15: Stress dependence of primary creep strain ε_{pr}^{cr} for the 9Cr-1Mo-V-Nb (ASTM P91) heat-resistant steel at 600°C in the complete stress range.

In many high-strength alloys this creep damage appears early in life and leads to failure after small creep strains (approximately 1%). In high-temperature design of structures it is important to make sure:

- that the *creep strain* ε^{cr} during the design life is acceptable;

- that the *creep ductility* ε^* (strain to failure) is adequate to cope with the acceptable creep strain;

- that the *time-to-rupture* t^* at the design loads σ and temperatures T is longer (by a suitable safety factor) than the design life.

Times-to-failure are normally presented as *creep-rupture* diagrams or *long-term strength* curves, as illustrated on Fig. 3.16. Their application is obvious: if you know

3. Non-isothermal creep-damage model for a wide stress range

the stress σ and the temperature T you can read off the life t^*; if you wish to design for a certain life t^* at a certain temperature T, you can read off the design stress σ.

Typical *long-term strength* curves for advanced heat-resistant steels (see e.g. Fig. 3.16) demonstrate the transition from ductile damage character dominant in the "high" stress range to the brittle damage character dominant in the "low" stress range, i.e. mixed damage mode during the "moderate" stress range, see e.g. [25, 152, 166, 181]. The brittle damage character dominant in the "low" stress range is caused the overaging and material microstructure degradation for the advanced heat-resistant steels. Thus, brittle damage character is accompanied by the significant decreasing of the *time-to-rupture* t^*, as was shown in experimental studies e.g. [5, 71, 102, 103, 109, 122, 155]. Simple extrapolation (e.g. with Larson-Miller parameter) of *long-term strength* caused by ductile damage at "high" stresses may lead to considerable overestimation of *time-to-rupture* t^* at "moderate" and "low" stress ranges. Long-term creep-rupture data [64, 71, 102, 103, 118] obtained from creep tests longer than 50 000 h, show a change in the damage mechanism for longer failure times leading to premature failure in comparison with Larson-Miller parameter predictions, as shown in Fig. 3.16.

All the metallurgical changes occurring under creep conditions described in [71, 190] are of great importance in advanced creep-resistant steels because they strongly affect creep and failure properties. The creep-rupture testing results by the European and Japanese research organizations show that, especially for new advanced steels in the class of 9-12%Cr, long-term creep-rupture testing (over 50 000 h) under low stresses is required to determine the real alloy behaviour, e.g. refer to [101, 118]. Otherwise a very dangerous overestimation of the material stability can result as shown in different publications if comparing long term extrapolations on the basis of either short or long term tests. A gradual loss in a long-term creep-rupture strength is caused by the changes of material microstructure during thermal exposure and creep deformations and is proved by the change of the slope of *long-term strength* slope with a decrease in stress, see e.g. [5, 102, 103, 118, 122]. Therefore, thermal ageing effects and the loss of creep-rupture strength must be taken into account for the formulation of constitutive equations of the long-term strength model in the form of ductile to brittle damage character transition.

3.3. Non-isothermal long-term strength and tertiary creep modeling

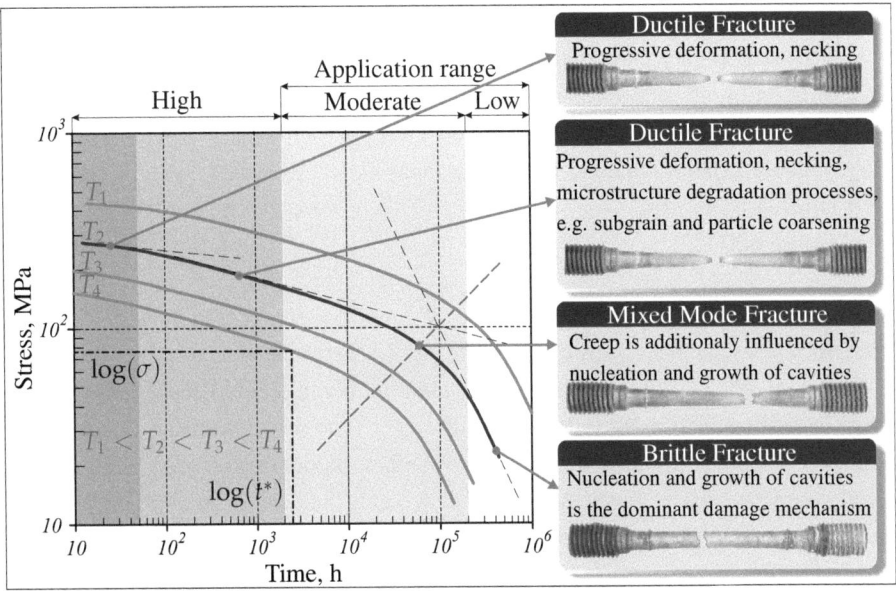

Figure 3.16: Schematic *long-term strength* curves at various temperatures T for advanced heat-resistant steels illustrating the transition of damage character, after [25, 152, 166, 181].

3.3.1 Stress dependence

An important problem of the creep-damage modeling is the ability to extrapolate the laboratory creep-rupture data usually obtained under increased stress σ and temperature T to the in-service loading conditions in "moderate" and "low" ranges of stress σ and temperature T. The tertiary creep parameters corresponding to "low" stress range can be identified only from the extrapolation of *long-term strength* curve based on the assumptions of ductile to brittle damage character transition [166], and accelerated microstructure degradation [5, 102] during thermal exposure and creep deformations.

All the available experimental creep-rupture data from [44, 102, 107, 109, 118, 162, 180, 188, 189] for the 9Cr-1Mo-V-Nb (ASTM P91) heat-resistant steel at 600°C illustrated on Fig. 3.17 shows that only "high" and "moderate" stress ranges can be analysed basing on experiments. Thereby, it is possible to formulate the part of

3. Non-isothermal creep-damage model for a wide stress range

long-term strength curve fitting the "high" stress experimental creep-rupture data and to identify the tertiary creep parameters corresponding to ductile damage character. The formulation of *long-term strength* equation for the complete stress range of the 9Cr-1Mo-V-Nb (ASTM P91) steel at 600°C begins with the "high" stress range. The available experimental creep-rupture data [44, 102, 107, 109, 118, 162, 180, 188, 189] is fitted by the Eq. (3.3.17) for "linear" *long-term strength* curve at constant temperature 600°C:

$$t^* = b_2\, \sigma^{-(n-k)}, \qquad (3.3.18)$$

where $n = 12$ is the power-law exponent from (3.1.2) for the 9Cr-1Mo-V-Nb (ASTM P91) steel at 600°C, and the tertiary creep material parameter b_2 for the "high" stress range is defined as $b_2 = 1.25 \cdot 10^{28}\,[\text{h} \cdot \text{MPa}^{(n-k)}]$. The value of tertiary material parameter $k = 0.5$ defines the difference in power-law exponents between the minimum creep strain rate $\dot{\varepsilon}^{cr}_{min}$ vs. stress σ dependence and the *long-term strength* curve in the complete stress range. And referring to [152] tertiary creep parameter k must have the value lying in range $(0 < k < 1)$ for the creep-damage models based on Kachanov-Rabotnov concept [88, 100, 173].

Unfortunately, there is no creep-rupture experimental data in the "low" stress range available in literature for the 9Cr-1Mo-V-Nb (ASTM P91) steel at 600°C. It is caused by the extremely long duration of necessary creep tests which should be over 100 000 hours to achieve the critical phase of creep rupture for low stress values less than 70 MPa. Visual observation of experimental creep-rupture data [44, 102, 107, 109, 118, 162, 180, 188, 189] for the 9Cr-1Mo-V-Nb (ASTM P91) steel at 600°C shows that the approximate location of transition from "ductile" to "brittle" damage character can be detected in the "moderate" stress range. This transition is visualized in the graduate change of slope of long-term strength curve with the reduction of stress σ and is proved by the available creep-rupture experimental data in the "moderate" stress range, as illustrated on Fig. 3.17. Therefore, the expected location of the "linear" *long-term strength* curve in "low" stress range can be estimated using the approximate value of the transition stress $\tilde{\sigma}_0$ from ductile to brittle damage character. Since the approximate value of the transition stress is taken as $\tilde{\sigma}_0 = 100$ MPa, than the equation for the *long-term strength* curve at constant temperature 600°C is following:

$$t^* = b_1\, \sigma^{-(1-k)}, \qquad (3.3.19)$$

3.3. Non-isothermal long-term strength and tertiary creep modeling

where the tertiary creep material parameter b_1 for the "low" stress range is identified as $b_1 = 1.25 \cdot 10^6 \ [\text{h} \cdot \text{MPa}^{(1-k)}]$ by the manual fitting.

The value of stress $\tilde{\sigma}_0$, which denotes the transition from ductile to brittle damage mechanism at 600°C, have been taken the same as the stress σ_0, which denotes transition from viscous to power-low creep at 600°C in Eq. (3.1.2). But for other temperatures the values of $\tilde{\sigma}_0$ and σ_0 may be rather different. Using the defined value $\tilde{\sigma}_0 = 100$ MPa it is possible to formulate the following equation of the *long-term strength* curve in the complete stress range with previously identified tertiary creep material parameters:

$$t^* = \left(\frac{\sigma^{1-k}}{b_1} + \frac{\sigma^{n-k}}{b_2} \right)^{-1} = \frac{B}{\sigma^{1-k}\tilde{\sigma}_0^{n-1} + \sigma^{n-k}}, \quad (3.3.20)$$

where the tertiary creep material parameter B for the complete stress range is identified as $B = b_2 = 1.25 \cdot 10^{28} \ [\text{h} \cdot \text{MPa}^{(n-k)}]$. The *long-term strength* curve (3.3.20) illustrated on Fig. 3.17 takes into account ductile damage mode for "high" stresses, brittle damage mode for "low" stresses and mixed damage mode for "moderate" stresses, defined by the transition stress value $\tilde{\sigma}_0 = 100$ MPa.

3.3.2 Creep-damage coupling

The next step in the phenomenological approach to creep modeling is the formulation of damage mechanism incorporated into the creep constitutive model (3.1.2). To characterize the damage processes the creep constitutive equation (3.1.2) must be generalized by introduction of the damage internal state variables and appropriate evolution equations. Isotropic damage models are generally formulated using the the concepts of scalar *damage parameter* ω using Eq. (1.4.13) and the *effective stress* $\tilde{\sigma}$ using Eq. (1.4.15) as described in Sect. 1.4, see e.g. [34, 124, 152, 166].

In the frames of the developed creep-damage model it is necessary to introduce the damage parameter ω which reflects both ductile and brittle damage characters. Due to the Kachanov-Rabotnov concept [100, 173] the value of damage parameter ω lies in the range $(0 \leq \omega \leq 1)$, where the $\omega = 0$ corresponds to the undamaged state in the initial moment of time t and $\omega = 1$ corresponds to fracture when $t = t^*$. The accumulation character of damage parameter ω can be analyzed using the following

3. Non-isothermal creep-damage model for a wide stress range

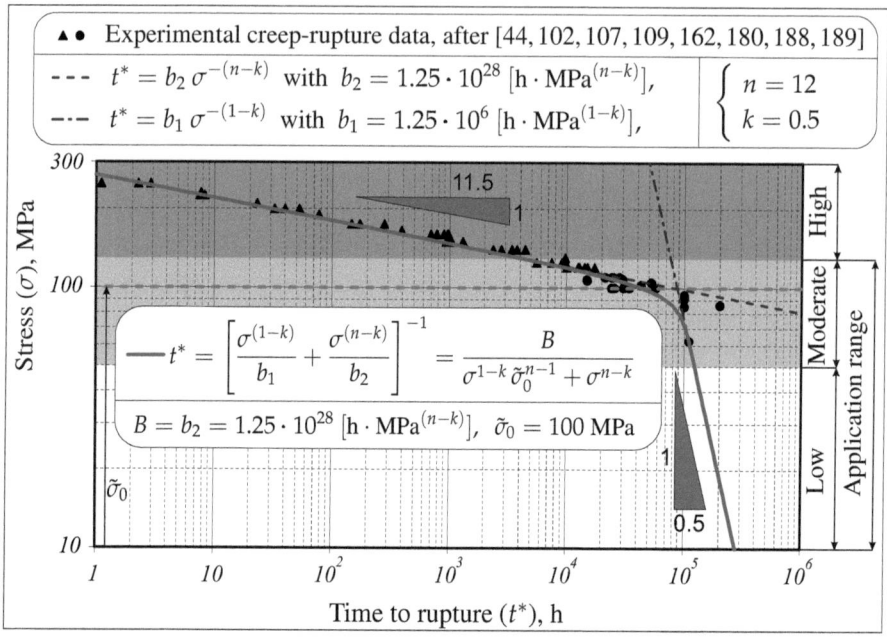

Figure 3.17: Experimental creep-rupture data [44, 102, 107, 109, 162, 180, 188, 189] for the 9Cr-1Mo-V-Nb (ASTM P91) steel at 600°C fitted by the Eq. (3.3.20) for *long-term strength* curve.

equation, as illustrated on Fig. 3.18:

$$\omega(t) = 1 - \left(1 - \frac{t}{t^*}\right)^{\frac{1}{l+1}}, \qquad (3.3.21)$$

where the tertiary creep material constant l governs the accumulation character of damage parameter ω and defines the fracture ductility ε^*. The damage character can be approximately estimated using the values of l considering the following criterion: $l < 5$ corresponds to ductile damage mode, $l > 15$ corresponds to brittle damage mode, and ($5 \leq l \leq 15$) describes the "mixed" damage mode, as illustrated on Fig. 3.18.

Creep-rupture observations for 9-12%Cr heat-resistant steels in [122] shows that

- "breakdown of creep strength" (rapid decrease of the slope of *long-term*

3.3. Non-isothermal long-term strength and tertiary creep modeling

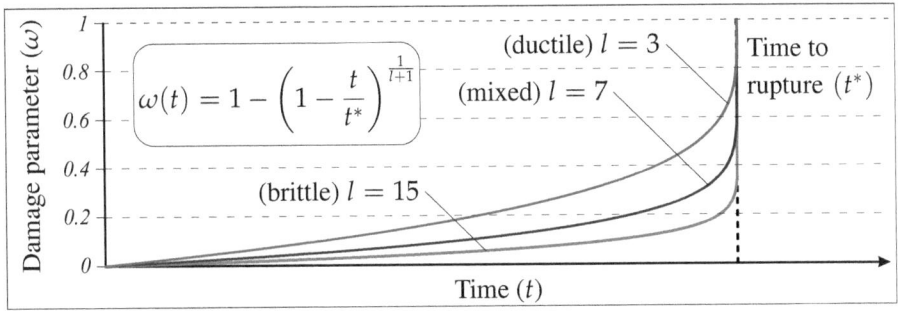

Figure 3.18: Classification of accumulation character types for the scalar damage parameter ω using Eq. (3.3.21) according to Kachanov-Rabotnov concept [88, 100, 173].

strength curve and decrease of creep ductility ε^*) is connected with brittle damage mode,

- brittle damage mode is only dominant in the linear creep range under low stress level,

- ductile damage processes influence on the creep deformations in the power law creep range.

Therefore, by analogy to Kachanov-Rabotnov-Hayhurst creep-damage model [88, 100, 173] a new creep constitutive equation (3.2.11) based on double power-law stress response function (3.1.2) is formulated using the *strain equivalence principle* after Lemaitre [123] and ductile to brittle damage character transition [122, 155] as follows

$$\dot{\varepsilon}^{cr} = A\frac{\sigma H}{1-\omega_b} + \frac{A}{\sigma_0^{n-1}}\left(\frac{\sigma H}{1-\omega_d}\right)^n, \qquad (3.3.22)$$

including the strain-hardening function (3.2.12) to describe the primary creep stage

$$H(\varepsilon^{cr}) = 1 + \alpha\ \exp(-\beta\ \varepsilon^{cr}). \qquad (3.3.23)$$

In notation (3.3.22) the values of secondary creep parameters defined in Eq. (3.1.2) for the 9Cr-1Mo-V-Nb heat-resistant steel at 600°C are following: $n = 12$, $A = 2.5 \cdot 10^{-9}$ [MPa^{-1}/h] and the transition stress $\sigma_0 = 100$ MPa. The variable ω_b denotes the brittle damage parameter and variable ω_d denotes the ductile damage parameter. In notation (3.3.23) the values of primary creep parameters defined in Sect. 3.2.2 for the 9Cr-1Mo-V-Nb (ASTM P91) heat-resistant steel at 600°C are following: $\alpha = 0.5$ and $\beta = 300$.

Creep constitutive equation (3.3.22) must be accompanied with the damage evolution equations. They were formulated for the brittle ω_b and ductile ω_d damage parameters similarly to Kachanov-Rabotnov-Hayhurst creep-damage model [88, 100, 173] as follows

$$\dot{\omega}_b = \frac{1}{t^*(l_1+1)(1-\omega_b)^{l_1}} \quad \text{and} \quad \dot{\omega}_d = \frac{1}{t^*(l_2+1)(1-\omega_d)^{l_2}}, \quad (3.3.24)$$

where $t^*(\sigma)$ is the stress-dependent time-to-rupture function according to Eq. (3.3.20), and the tertiary creep material parameters for the 9Cr-1Mo-V-Nb (ASTM P91) heat-resistant steel at 600°C defined by Eq. (3.3.20) are the following: $B = 1.25 \cdot 10^{28}$ [h · MPa$^{(n-k)}$], $k = 0.5$ and transition stress $\tilde{\sigma}_0 = 100$ MPa. In Eqs (3.3.20) and (3.3.24) the power-law exponent $n = 12$ is taken from the constitutive equation (3.3.22). The tertiary creep material parameters l_1 governing the brittle damage mode and l_2 governing the ductile damage mode have to be defined from the experimental data for creep-fracture ductility ε^*.

Formulating the damage accumulation equation (3.3.21) for the brittle and ductile values of creep parameter l one can obtain the following mathematical conversion

$$\left(1 - \frac{t}{t^*}\right) = (1-\omega_b)^{l_1+1} = (1-\omega_d)^{l_2+1}. \quad (3.3.25)$$

Hereby, in the proposed creep-damage model (3.3.22) - (3.3.24) only the one damage evolution equation can be formulated, if to replace one of the damage parameters (ω_b or ω_d) in the creep constitutive equation (3.3.22) by one of the following corresponding connections:

$$\omega_b = 1 - (1-\omega_d)^{\frac{l_2+1}{l_1+1}} \quad \text{or} \quad \omega_d = 1 - (1-\omega_b)^{\frac{l_1+1}{l_2+1}}. \quad (3.3.26)$$

For a constant stress $\sigma = $ const both of the damage evolution equations (3.3.24) can be integrated by time t assuming the integration limits as $\omega_b = \omega_d = 0$ corresponding to $t = 0$ and $\omega_b = \omega_d = 1$ corresponding to $t = t^*$. As a result one can obtain

3.3. Non-isothermal long-term strength and tertiary creep modeling

the *long-term strength* equation (3.3.20). Integration of the both of damage evolution equations (3.3.24) taking into account the *long-term strength* equation (3.3.20) provides both the damage parameters $\omega_b(t)$ and $\omega_d(t)$ as a functions of time defined by Eq. (3.3.21).

Taking into account Eq. (3.3.21) and neglecting the strain-hardening function (3.3.23), the creep strain rate equation (3.3.22) can be integrated analytically by time t leading to the following creep strain ε^{cr} vs. time t dependence:

$$\varepsilon^{cr}(t) = A\sigma \frac{l_1+1}{l_1} t^* \left[1 - \left(1 - \frac{t}{t^*}\right)^{\frac{l_1}{l_1+1}} \right] + \\ + \frac{A}{\sigma_0^{n-1}} \sigma^n \frac{l_2+1}{l_2+1-n} t^* \left[1 - \left(1 - \frac{t}{t^*}\right)^{\frac{l_2+1-n}{l_2+1}} \right], \quad (3.3.27)$$

From Eq. (3.3.27) it follows that the constant l_2 must satisfy the condition ($l_2 > n - 1$) providing the positive creep strain ε^{cr} for the positive stress σ values. By setting $t = t^*$ in the form of Eq. (3.3.20) the creep strain accumulated during the secondary and tertiary stages before the fracture, i.e. $\varepsilon^{cr}_{s/t} = \varepsilon^{cr}(t^*)$, can be calculated in the form of dependence on stress as

$$\varepsilon^{cr}_{s/t}(\sigma) = \frac{B}{\sigma^{1-k}\tilde{\sigma}_0^{n-1} + \sigma^{n-k}} \left(A\sigma \frac{l_1+1}{l_1} + \frac{A}{\sigma_0^{n-1}}\sigma^n \frac{l_2+1}{l_2+1-n} \right). \quad (3.3.28)$$

It can be observed from Eq. (3.3.28) that $\varepsilon^{cr}_{s/t} \sim \sigma^k$, where the value of tertiary material parameter $k = 0.5$ was considered in *long-term strength* equations (3.3.19) - (3.3.20). As it was mentioned before k defines the difference in power-law exponents between the minimum creep strain rate $\dot{\varepsilon}^{cr}_{min}$ vs. stress σ dependence and the *long-term strength* curve in the complete stress range. For the positive values of k the fracture strain ε^* increases with an increase in the stress σ value due to notation $\varepsilon^* \sim \sigma^k$. Such a dependence is usually observed for many alloys in the case of moderate stresses, as mentioned in [152]. Therefore, the experimental creep fracture strain ε^* values [102, 169, 188, 201] available for "high" and "moderate" stress levels can be fitted by the following approximation as illustrated on Fig. 3.19:

$$\varepsilon^*(\sigma) = \varepsilon^{cr}_{pr}(\sigma) + \varepsilon^{cr}_{s/t}(\sigma) \quad \text{with} \quad \varepsilon^{cr}_{s/t}(\sigma) = C\sigma^k, \quad (3.3.29)$$

3. Non-isothermal creep-damage model for a wide stress range

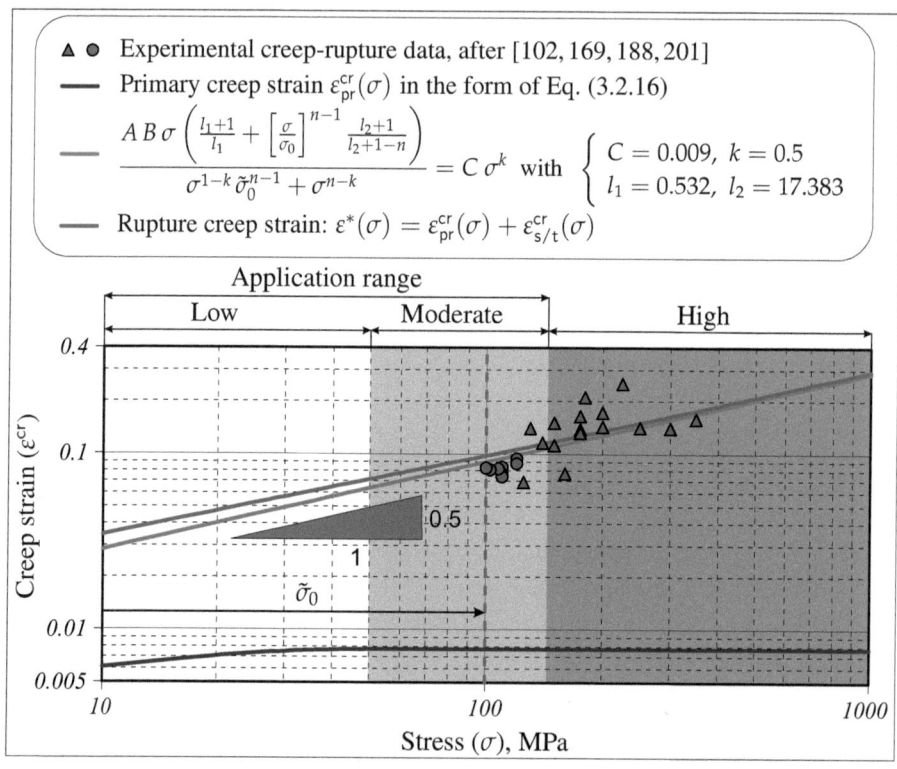

Figure 3.19: Experimental data [102, 169, 188, 201] presenting rupture creep strain ε^* vs. stress σ dependence fitted by the proposed creep-damage model (3.3.22) - (3.3.24) for the 9Cr-1Mo-V-Nb (ASTM P91) heat-resistant steel at 600°C.

where $\varepsilon_{pr}^{cr}(\sigma)$ denotes the primary creep strain defined by Eq. (3.2.16), $\varepsilon_{s/t}^{cr}(\sigma)$ is the creep strain accumulated during the secondary and tertiary stages before the fracture defined by Eq. (3.3.28), and the values of fitting parameters are defined as $k = 0.5$ and $C = 0.009$.

The effect of brittle damage character presented by parameter ω_b is dominant only in linear (viscous) creep range. Than the value of tertiary creep parameter $l_1 = 0.532$ defining brittle damage mode can be defined from the "linear-brittle" part of

3.3. Non-isothermal long-term strength and tertiary creep modeling

Eq. (3.3.28) as follows:

$$\frac{A B \sigma^k (l_1 + 1)}{\tilde{\sigma}_0^{n-1} l_1} = C \sigma^k. \qquad (3.3.30)$$

The effect of ductile damage character presented by parameter ω_d is dominant only in power-law creep range. Than the value of tertiary creep parameter $l_2 = 17.383$ defining ductile damage mode can be defined from the "power-law-ductile" part of Eq. (3.3.28) as follows:

$$\frac{A B \sigma^k (l_2 + 1)}{\tilde{\sigma}_0^{n-1} (l_2 + 1 - n)} = C \sigma^k. \qquad (3.3.31)$$

Numerical integration by time t of the both creep constitutive equation (3.3.22) and damage evolution equations (3.3.24) with the constant stress σ values and all previously defined creep material parameters (A, n, σ_0, B, k, $\tilde{\sigma}_0$, l_1, l_2) produce the creep strain ε^{cr} vs. time t dependence or uniaxial *creep curve*. Using the creep-damage model (3.3.22) - (3.3.24) one can obtain the valid uniaxial *creep curves* in the stress range from 0 MPa till 300 MPa at temperature 600°C. The illustrated on Fig. 3.20 creep curves for "moderate" and "low" stress range obtain by model (3.3.22) - (3.3.24) show the *locus of elongation at failure*. That locus demonstrates the growth of creep-rupture strain ε^* and the transition from brittle to ductile damage character with the growth of stress σ.

Creep curves obtained by the creep-damage model (3.3.22) - (3.3.24) can be compared with an available experimental creep curves [165, 188] for "high" stress range for the 9Cr-1Mo-V-Nb (ASTM P91) heat-resistant steel at 600°C, as illustrated on Fig. 3.21. Some disagreement with available experimental curves from [165, 188] can be explained by the considerable variation of experimental results. Thus, the inaccuracy of creep experimental measurements can reach till 30%. But the character of creep behaviour at primary, secondary and tertiary stages and the critical failure parameters (time-to-rupture t^* and rupture strain t^*) for the creep-damage model (3.3.22) - (3.3.24) and experimental creep creep curves [165, 188] are quite close.

It should be taken into account that experimental data may show a large scatter generated by testing a series of specimens removed from the same material. The origins of scatter in creep testing are discussed in [60]. Furthermore, unlike the small strain elasticity, the creep behavior may significantly depend on the kind of processing of specimens, e.g. the heat treatment. As a result, different data sets for

3. Non-isothermal creep-damage model for a wide stress range

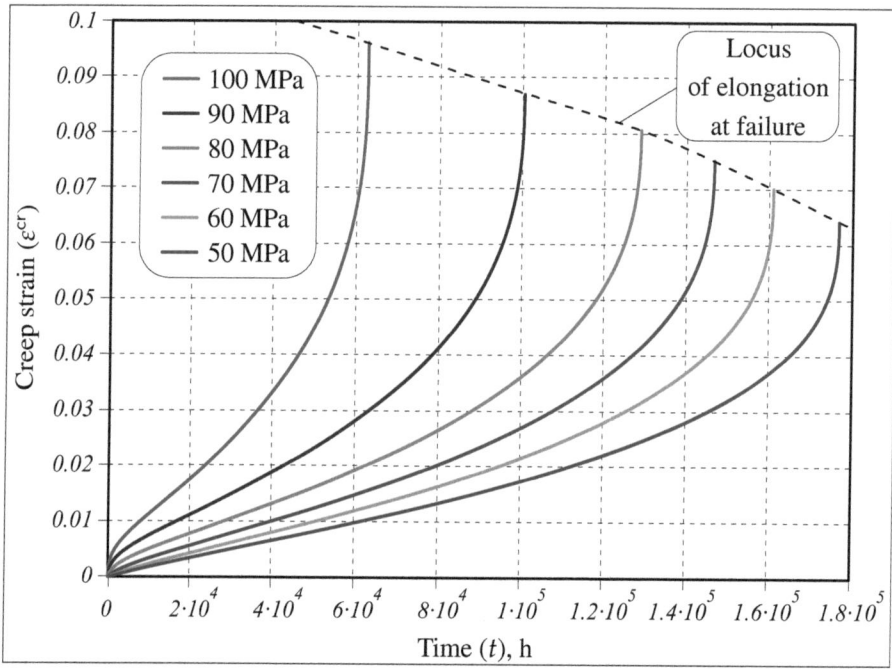

Figure 3.20: Creep curves obtained by the numerical integration of creep-damage model (3.3.22) - (3.3.24) by time for the values of tensile stress σ correspondent to "low" and "moderate" stress ranges for the 9Cr-1Mo-V-Nb (ASTM P91) heat-resistant steel at 600°C.

the material with the same chemical composition may be found in the literature. For example, one may compare experimental data for the 9Cr-1Mo-V-Nb (ASTM P91) ferritic steel obtained in different laboratories [4, 48, 65, 110, 165, 201].

3.3.3 Temperature dependence

Figure 3.22 illustrates all the available creep-rupture experimental data for the 9Cr-1Mo-V-Nb (ASTM P91) heat-resistant steel at different temperatures and shows the necessity to analyze the damage process for the "low" and "high" stress regions separately for the purpose of material parameter identification. An important problem of the creep constitutive modeling is the ability to extrapolate the laboratory

3.3. Non-isothermal long-term strength and tertiary creep modeling

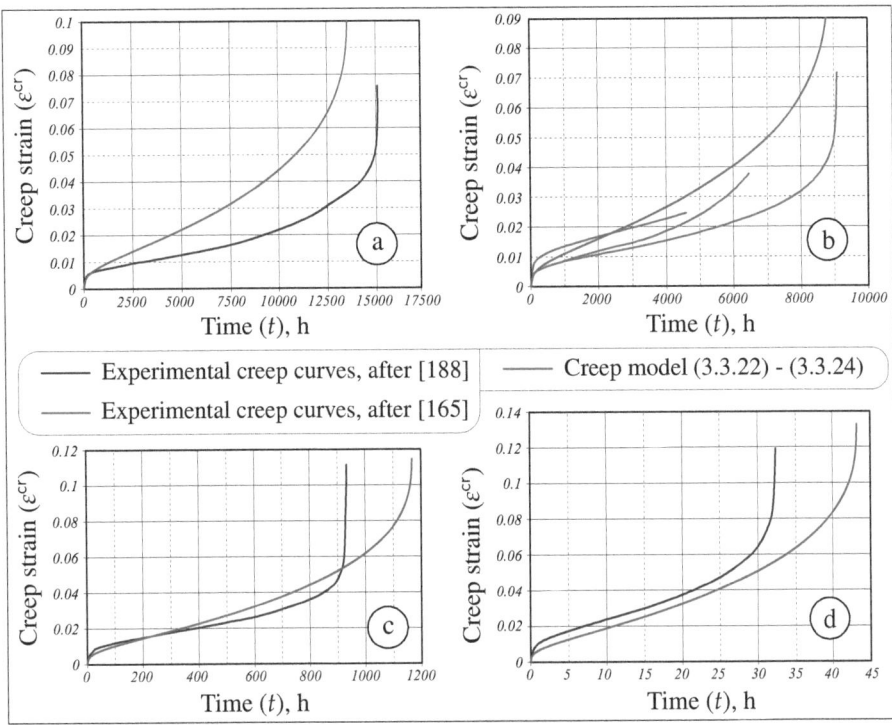

Figure 3.21: Creep curves obtained by creep-damage model (3.3.22) - (3.3.24) comparing to experiments [165, 188] for the 9Cr-1Mo-V-Nb (ASTM P91) heat-resistant steel at 600°C for the following values of tensile stress σ: a) 120 MPa, b) 125 MPa, c) 150 MPa, d) 200 MPa.

creep-rupture data usually obtained under increased stress σ and temperature T to the in-service loading conditions, for which the experimental data is not available. Since the damage phenomenon is also a temperature activated process, the proposed time-to-rupture t^* on stress σ isothermal dependence (3.3.20) is extended to the case of various temperatures by the introduction of temperature-dependent tertiary creep constants. And it provides the satisfactory fit of available creep-rupture experimental data [32, 44, 71, 72, 102, 107, 109, 162, 180, 188, 189] at least for the four different temperatures, as illustrated on Fig. 3.22.

Thereby, Eq. (3.3.20) is transformed into the set of non-isothermal long-term

3. Non-isothermal creep-damage model for a wide stress range

strength curves (see Fig. 3.22) by the introduction of two Arrhenius-type functions in the following way

$$t^*(T) = \frac{B(T)}{\sigma^{1-k}\,[\tilde{\sigma}_0(T)]^{n-1} + \sigma^{n-k}}, \quad (3.3.32)$$

where $B(T)$ denotes the temperature-dependent creep-rupture parameter

$$B(T) = B_f \exp\left(\frac{Q_f}{RT}\right), \quad (3.3.33)$$

and $\tilde{\sigma}_0(T)$ is the temperature-dependent stress, which denotes the transition from ductile to brittle damage mode

$$\tilde{\sigma}_0(T) = B_\sigma \exp\left(\frac{\tilde{Q}_\sigma}{RT}\right). \quad (3.3.34)$$

In Eqs (3.3.33) - (3.3.34) the dependence on temperature is introduced by the Arrhenius-type functions with the following tertiary creep material parameters: $B_f = 2.0 \cdot 10^{-14}$ [h $\cdot$ MPa$^{(n-k)}$], $Q_f = 698000$ [J $\cdot$ mol^{-1}], $B_\sigma = 0.205$ MPa, $\tilde{Q}_\sigma = 44940$ [J $\cdot$ mol^{-1}] and universal gas constant $R = 8.314$ [J $\cdot$ K^{-1} $\cdot$ mol^{-1}].

The isothermal creep-damage model (3.3.22) - (3.3.24) is transformed into the non-isothermal form consisting of the creep constitutive equation (3.3.35) for the creep strain rate and two damage evolution equations (3.3.37) - (3.3.38) for the damage accumulation rate. The non-isothermal creep constitutive equation describing primary and steady-state creep behaviour is formulated analogously to Eq. (3.3.22) as follows

$$\dot{\varepsilon}^{cr} = A(T)\frac{\sigma\left[1+\alpha\,\exp\left(-\beta\,\varepsilon^{cr}\right)\right]}{1-\omega_b} + \frac{A(T)}{[\sigma_0(T)]^{n-1}}\left(\frac{\sigma\left[1+\alpha\,\exp\left(-\beta\,\varepsilon^{cr}\right)\right]}{1-\omega_d}\right)^n, \quad (3.3.35)$$

where the temperature-dependent secondary creep parameters $A(T)$ and $\sigma_0(T)$ are formulated using the Arrhenius-type functions (3.1.5) and (3.1.8) as follows

$$\begin{aligned}A(T) &= A_1(T) = A_c \exp\left(\frac{-Q_c}{RT}\right)\\ \text{and} \quad \sigma_0(T) &= A_\sigma \exp\left(\frac{-Q_\sigma}{RT}\right).\end{aligned} \quad (3.3.36)$$

3.3. Non-isothermal long-term strength and tertiary creep modeling

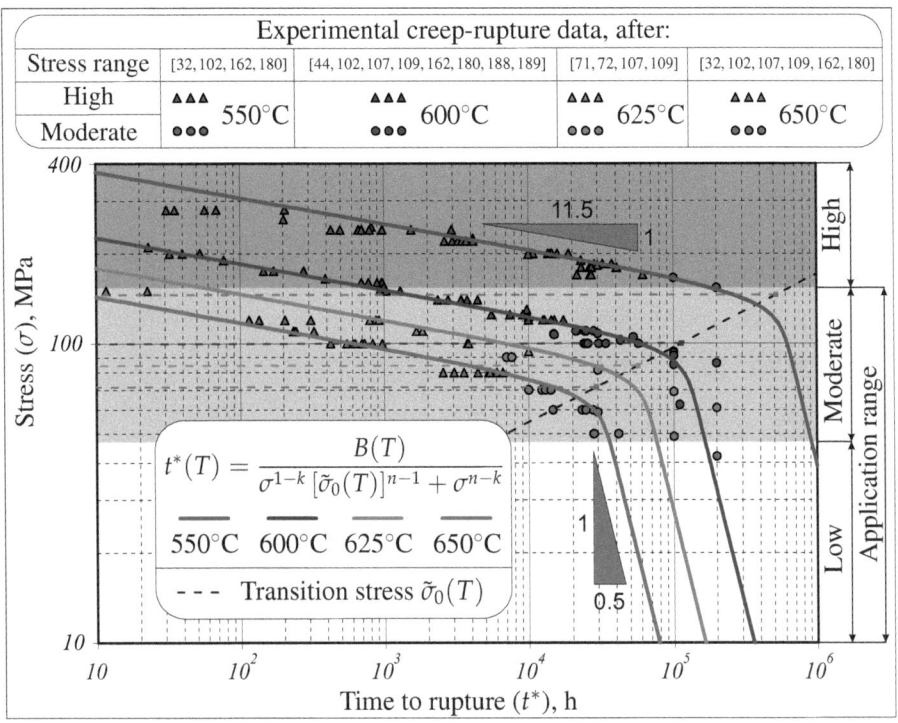

Figure 3.22: Experimental creep-rupture data [32, 44, 71, 72, 102, 107, 109, 162, 180, 188, 189] at various temperatures for the 9Cr-1Mo-V-Nb (ASTM P91) heat-resistant steel fitted by the proposed non-isothermal Eq. (3.3.32) for *long-term strength* curves.

In notations (3.3.35) - (3.3.36) the values of secondary creep parameters for the 9Cr-1Mo-V-Nb (ASTM P91) steel defined in Sect. 3.1 are following: $n = 12$, $A_c = A_{01} = 2300$ [MPa^{-1}/h], $Q_c = Q_1 = 200000$ [J · mol^{-1}], $A_\sigma = 0.658$ MPa, $Q_\sigma = 36364$ [J · mol^{-1}] and the universal gas constant $R = 8.314$ [J · K^{-1} · mol^{-1}]. The values of primary creep parameters in Eq. (3.3.35) defined in Sect. 3.2 are following: $\alpha = 0.5$ and $\beta = 300$.

Taking into account the time-to-rupture function (3.3.32), the non-isothermal damage evolution equations describing tertiary creep behaviour and fracture are formulated analogously to Eq. (3.3.24) for the brittle damage parameter ω_b

$$\dot{\omega}_b = \frac{\sigma^{1-k}[\tilde{\sigma}_0(T)]^{n-1} + \sigma^{n-k}}{B(T)(l_1+1)(1-\omega_b)^{l_1}} \qquad (3.3.37)$$

and the ductile damage parameter ω_d

$$\dot{\omega}_d = \frac{\sigma^{1-k}[\tilde{\sigma}_0(T)]^{n-1} + \sigma^{n-k}}{B(T)(l_2+1)(1-\omega_d)^{l_2}}, \qquad (3.3.38)$$

where the temperature-dependent tertiary creep parameters $B(T)$ and $\tilde{\sigma}_0(T)$ are formulated using the Arrhenius-type functions as follows

$$B(T) = B_f \exp\left(\frac{Q_f}{RT}\right) \quad \text{and} \quad \tilde{\sigma}_0(T) = B_\sigma \exp\left(\frac{\tilde{Q}_\sigma}{RT}\right). \qquad (3.3.39)$$

In notations (3.3.37) - (3.3.39) the values of tertiary creep material parameters are $l_1 = 0.532$, $l_2 = 17.383$, $B_f = 2.0 \cdot 10^{-14}$ [h · MPa$^{(n-k)}$], $Q_f = 698000$ [J · mol^{-1}], $B_\sigma = 0.205$ MPa, $\tilde{Q}_\sigma = 44940$ [J · mol^{-1}] and the universal gas constant $R = 8.314$ [J · K^{-1} · mol^{-1}].

Numerical integration by time of the both creep constitutive equation (3.3.35) and damage evolution equations (3.3.37) - (3.3.38) with the defined values of temperature T and stress σ produce the creep strain ε^{cr} vs. time t dependence or uniaxial *creep curve*. Using the non-isothermal creep-damage model (3.3.35) - (3.3.39) one can obtain the valid uniaxial *creep curves* in the stress range from 0 MPa till 300 MPa and temperature range from 550°C till 650°C.

3.4 Stress-dependent failure criterion

Multi-axial stress state has several effects on damage accumulation, see e.g. [152, 166]. It determines the parameter used to correlate the damage rate under different types of stress state, as to whether the maximum tensile stress $\sigma_{\max t}$, the von Mises effective stress σ_{vM}, or some other measure should be used. In addition, multiaxial constraint affects the ductility. The higher the level of constraint, the lower the ductility. The accumulation of creep and fatigue damage under multiaxial stress are both important factors as far as evaluating the life of engineering components is concerned. For this reason, they have been in the focus of considerable interest for many

3.4. Stress-dependent failure criterion

years. And some relevant highlights of the developments arising from this interest have to be discussed below.

In the early 1960s, Johnson, Henderson and Khan [99] built on work done by Sdobyrev [182] to characterize the stress dependence of creep continuum damage. The model chosen is based on the assumption that creep damage is stress dependent, and that two stress related parameters are relevant, the von Mises effective stress σ_{vM} and the maximum principal (tensile) stress $\sigma_{max\,t}$. To analyze the failure mechanisms under multi-axial stress state and high-temperature creep conditions the *isochronous rupture loci* are conventionally used. They illustrate stress states leading to the same *time-to-fracture t^**. The general model which correlates damage under uniaxial tension with damage under more complex conditions is introduced by adopting the damage equivalent stress σ_{eq}^{ω}. Therefore, the early approaches to long-term strength assessments are based on the damage equivalent stress σ_{eq}^{ω}, which is governed either by the maximum tensile stress $\sigma_{max\,t}$ or by the von Mises effective stress σ_{vM}, see Fig. 3.23a.

Later Hayhurst [85, 96, 121] proposed the damage equivalent stress σ_{eq}^{ω} as a linear combination of the maximum tensile stress $\sigma_{max\,t}$ and the von Mises effective stress σ_{vM} as follows:

$$\sigma_{eq}^{\omega} = \alpha\, \sigma_{max\,t} + (1 - \alpha)\, \sigma_{vM} \tag{3.4.40}$$

with the maximum tensile stress in form

$$\sigma_{max\,t} = (\sigma_I + |\sigma_I|)/2 \tag{3.4.41}$$

and the von Mises effective stress in form

$$\sigma_{vM} = \sqrt{\frac{3}{2}\mathbf{s}\cdot\cdot\mathbf{s}}. \tag{3.4.42}$$

In notations (3.4.40) - (3.4.41) σ_I denotes the first principal stress, and α is a weighting factor considering the influence of damage mechanisms (σ_I-controlled or σ_{vM}-controlled). Figure 3.23a shows the failure criterion using the damage equivalent stress σ_{eq}^{ω} for two defined values of weighting factor $\alpha = 0.3$ and $\alpha = 0.5$. Such a form of the damage equivalent stress σ_{eq}^{ω} corresponds to a quite narrow ranges of stress σ and temperature T.

Generally, it has been assumed that the von Mises effective stress σ_{vM} controls creep rate and governs the nucleation of creep voids but does not contribute to their

3. Non-isothermal creep-damage model for a wide stress range

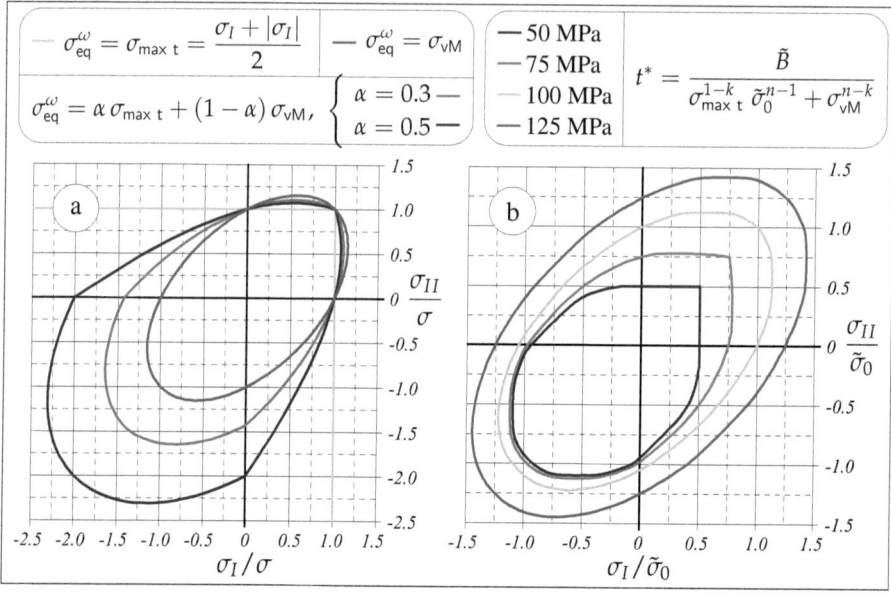

Figure 3.23: Formulation of multiaxial failure criteria using plane stress isochronous rupture loci: a) conventional approach, after [85, 96, 121], b) stress-dependent multiaxial failure criterion based on long-term strength of the 9Cr-1Mo-V-Nb (ASTM P91) heat-resistant steel at 600°C.

growth, while the mean stress $\bar{\sigma}$ in the following form

$$\bar{\sigma} = \frac{\sigma_I + \sigma_{II} + \sigma_{III}}{3} \qquad (3.4.43)$$

promotes the growth of the creep voids or cracks, refer to [155] for the details. In the torsional creep rupture tests [155] all of the specimens exhibited a ductile transgranular creep fracture of shear type. And larger creep deformations with many micro-voids were observed near the fracture surfaces. These results should be because the larger σ_{vM} and little $\bar{\sigma}$ existed in the torsional creep specimens. It has been reported [155] that at higher stresses (meanwhile larger $\bar{\sigma}$) a ductile fracture caused due to the larger deformation, which restrained the growth of creep voids. But at lower stresses a brittle intergranular fracture occurred due to the sufficient growth of voids with slower creep deformations. At each stress state, creep deformation and nucleation of creep voids in a specimen at a higher stress are easily to occur due to

3.4. Stress-dependent failure criterion

the larger σ_{vM}, but the growth of voids is restrained by the larger deformation. With decreasing the applied stress, a brittle intergranular fracture mode should be present due to the growth of creep voids. However, from tension to torsion the growth of creep voids becomes difficult due to the decrease of $\bar{\sigma}$. Therefore, it is suggested that in a specimen with existing of $\bar{\sigma}$ and smaller σ_{vM}, a brittle intergranular fracture should occur easily [155].

The new mixed stress-dependent failure criterion based on *creep-rupture* diagram (3.3.20) of the 9Cr-1Mo-V-Nb (ASTM P91) heat-resistant steel at 600°C was proposed for the multi-axial form of creep-damage model (3.3.22) - (3.3.24) as follows:

$$t^* = \frac{\tilde{B}}{\sigma_{max\,t}^{1-k} \tilde{\sigma}_0^{n-1} + \sigma_{vM}^{n-k}}. \qquad (3.4.44)$$

The failure criterion (3.4.44) includes both the first principal stress σ_I and the von Mises effective stress σ_{vM}, and therefore it is applicable for the complete stress range. Within the range of "high" stresses the creep damage is primarily ductile and leads to the necking of uniaxial specimen, so it is governed by the von Mises effective stress σ_{vM}. Within the range of "low" stresses the creep damage is primarily brittle and leads to the long-term degradation of material microstructure. In this case we assume that the creep damage is governed by the first principal stress σ_I. The "moderate" stress range is controlled by the both first principal stress σ_I and the von Mises effective stress σ_{vM}. The set of isochronous rupture loci presented on Fig. 3.23b illustrates the idea of failure mode transition from brittle rupture for "low" stresses to ductile rupture for "high" stresses.

The stress dependence of damage can be presented by *isochronous damage surface*, analogous to a yield surface [166]. The *isochronous damage surface* shown on Fig. 3.24 illustrates the alternative presentation of the failure criterion (3.4.44) for the multiaxial stress state. The *time-to-rupture* t^* is plotted as a function of principal stresses σ_I and σ_{II}. And it illustrates the plane stress conditions which lead to the same *time-to-rupture* t^* depending on range of stress. It shows that for the long test durations the creep failure is primarily determined the first principle stress σ_I. This assumption is confirmed by the multi-axial creep-rupture experiments for a lot of materials including advanced heat-resistant steels under different stress σ and temperature T conditions, see e.g. [87, 89, 155].

Finally the multiaxial form of the proposed creep-damage model (3.3.22) -

3. Non-isothermal creep-damage model for a wide stress range

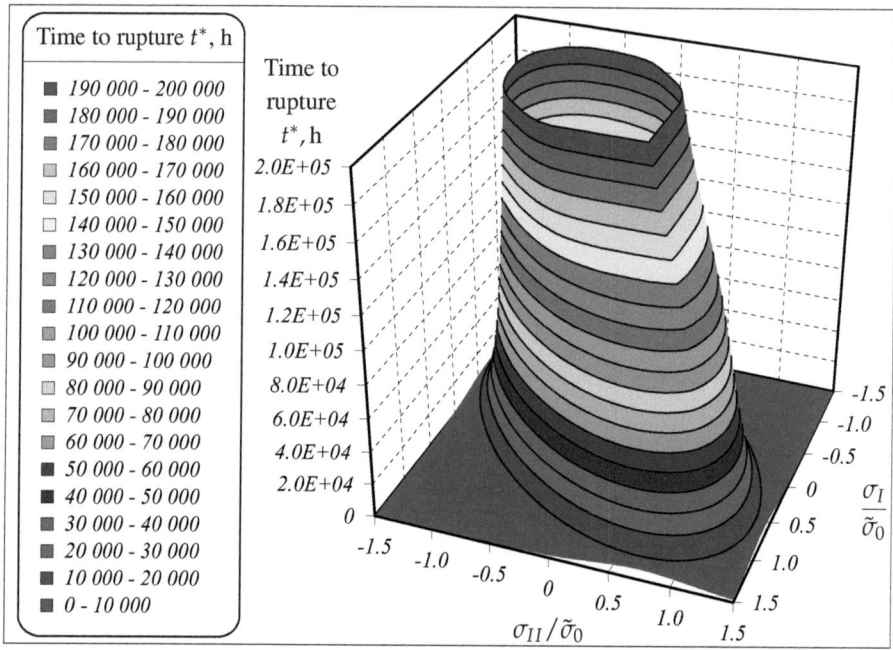

Figure 3.24: *Isochronous damage surface* presenting multiaxial stress-dependent failure criterion based on *long-term strength* of the 9Cr-1Mo-V-Nb (ASTM P91) heat-resistant steel at 600°C.

(3.3.24) is formulated consisting of constitutive equation for the creep strain rate tensor

$$\dot{\boldsymbol{\varepsilon}}^{\text{cr}} = \left[A \frac{\sigma_{\text{vM}} H}{1 - \omega_{\text{b}}} + \frac{A}{\sigma_0^{n-1}} \left(\frac{\sigma_{\text{vM}} H}{1 - \omega_{\text{d}}} \right)^n \right] \frac{3}{2} \frac{\boldsymbol{s}}{\sigma_{\text{vM}}} \quad (3.4.45)$$

with the strain-hardening function (3.2.12) depending on the equivalent creep strain $\varepsilon_{\text{eq}}^{\text{cr}}$

$$H(\varepsilon_{\text{eq}}^{\text{cr}}) = 1 + \alpha \, \exp(-\beta \, \varepsilon_{\text{eq}}^{\text{cr}}), \quad (3.4.46)$$

and two evolution equations for the brittle ω_{b} and ductile ω_{d} damage parameters

$$\dot{\omega}_{\text{b}} = [t^* (l_1 + 1)(1 - \omega_{\text{b}})^{l_1}]^{-1} \quad \text{and} \quad \dot{\omega}_{\text{d}} = [t^* (l_2 + 1)(1 - \omega_{\text{d}})^{l_2}]^{-1} \quad (3.4.47)$$

3.4. Stress-dependent failure criterion

with time-to-rupture function t^* depending on the stress parameters $\sigma_{\text{max t}}$ and σ_{vM}

$$t^*(\sigma_{\text{max t}}, \sigma_{\text{vM}}) = \frac{B}{\sigma_{\text{max t}}^{1-k} \tilde{\sigma}_0^{n-1} + \sigma_{\text{vM}}^{n-k}} \qquad (3.4.48)$$

where the maximum tensile stress $\sigma_{\text{max t}}$ is defined by Eq. (3.4.41), and the values of all previously identified creep material parameters for the 9Cr-1Mo-V-Nb (ASTM P91) heat-resistant steel at 600°C are following: $A = 2.5 \cdot 10^{-9}$ [MPa^{-1}/h], $n = 12$, $\sigma_0 = 100$ MPa, $\alpha = 0.5$, $\beta = 300$, $B = 1.25 \cdot 10^{28}$ [h · MPa$^{(n-k)}$], $\tilde{\sigma}_0 = 100$ MPa, $k = 0.5$, $l_1 = 0.532$, $l_2 = 17.383$.

It is necessary to notice that the time-to-rupture function $t^*(\sigma_{\text{max t}}, \sigma_{\text{vM}})$ in the form (3.4.48) for the multi-axial stress state can have different forms in the evolution equations (3.4.48). For instance, it can be dependent only on $\sigma_{\text{max t}}$ for evolution equation $\dot{\omega}_b$ or dependent only on σ_{vM} for evolution equation $\dot{\omega}_d$. The specific forms of time-to-rupture function t^* and the variants of stress parameters as it's arguments must be investigated and discussed in future.

For the case of the same time-to-rupture functions $t^*(\sigma_{\text{max t}}, \sigma_{\text{vM}})$ in the both evolution equations (3.4.47), the proposed creep-damage model (3.4.45) - (3.4.48) can be formulated with only one damage evolution equation, if to replace one of the damage parameters (ω_b or ω_d) in the creep constitutive equation (3.4.45) by one of the following corresponding connections, formulated analogously to Eq. (3.3.26):

$$\omega_b = 1 - (1 - \omega_d)^{\frac{l_2+1}{l_1+1}} \quad \text{or} \quad \omega_d = 1 - (1 - \omega_b)^{\frac{l_1+1}{l_2+1}}. \qquad (3.4.49)$$

Chapter 4

Creep estimations in structural analysis

In Chapters 2 and 3 we introduced creep constitutive and damage evolution equations for the modeling of creep in engineering materials. The objective of Chapt. 4 is the application of creep constitutive models to structural analysis. In Sect. 4.1 we start with the discussion of basic steps in modeling of creep in structures with the application of finite element method (FEM). The advantages of FEM application and creep analysis procedures in a commercial FEM-based software with conventional and user-defined constitutive models are highlighted. Section 4.2 is devoted to the introduction of benchmark problem concept in the frames of structural mechanics, description of its main purposes in creep mechanics and verification steps for a reliability assessment. As examples, two benchmark problems of creep-damage mechanics which can be solved by approximate numerical methods are presented. The reference solutions are compared with the finite element solutions by ANSYS and ABAQUS finite element codes with user-defined creep model subroutines. To discuss the applicability of the developed techniques to real engineering problems two examples of numerical life-time assessment with application of creep-damage models for initial-boundary value problems are presented in Sects 4.3 and 4.4. The purpose of Sect. 4.3 is the numerical creep behavior modeling of the T-piece pipe weldment made of the 9Cr-1Mo-V-Nb (ASTM P91) heat-resistant steel. The initially transversally-isotropic creep-damage model developed in Sect. 2.2.1 is incorporated in the commercial FEM-based CAE-software ANSYS. The numerical anal-

4. Creep estimations in structural analysis

ysis results of the welded structure qualitatively agree with available experimental observations and confirm the facts of various creep and damage material properties in longitudinal and transversal directions of a welding seam. Finally, in Sect. 4.4 an example of life-time assessment for the housing of a quick stop valve usually installed on steam turbines is presented. Long-term behavior of this components under approximately in-service loading conditions (constant internal pressure and constant temperature) is simulated by the FEM. The results show that the developed in Chapt. 3 constitutive model is capable to reproduce specific features of creep and damage processes in engineering structures during the long-term thermal exposure.

4.1   Application of FEM to creep-damage analysis

The aim of creep modeling is to reflect basic features of creep in engineering structures including the development of inelastic deformations, relaxation and redistribution of stresses as well as the local reduction of material strength. A model should be able to account for material deterioration processes in order to predict long-term structural behavior, to estimate the in-service life-time of a component and to analyze critical zones of failure caused by creep. Structural analysis under creep conditions usually requires the following steps, proposed in [147, 152]:

1. Assumptions must be made with regard to the geometry of the structure, types of loading and heating as well as kinematical constraints.

2. A suitable structural mechanics model (e.g. three-dimensional solids, beams, rods, plates and shells) must be applied based on the assumptions concerning kinematics of deformations, types of internal forces (moments) and related balance equations.

3. A reliable constitutive model must be formulated to reflect time dependent creep deformations and processes accompanying creep like hardening/recovery and damage.

4. A mathematical model of the structural behavior (initial-boundary value problem) must be formulated including the material independent equations, constitutive (evolution) equations as well as initial and boundary conditions.

4.1. Application of FEM to creep-damage analysis

5. Numerical solution procedures (e.g. the Ritz method, the Galerkin method, the finite element method) to solve non-linear initial-boundary value problems must be developed.

6. The verification of the applied models must be performed including the structural mechanics model, the constitutive model, the mathematical model as well as the numerical methods and algorithms.

For the numerical solution the direct variational methods, e.g. the Ritz method, the Galerkin method and the finite element method (FEM), are usually applied. In recent years the FEM has become the widely accepted tool for structural analysis. The advantage of the FEM is the possibility to model and analyze engineering structures with complex geometries, various types of loadings and boundary conditions. General purpose finite element codes ABAQUS, ANSYS, NASTRAN, COSMOS, MARC, ADINA, etc. were developed to solve various problems in solid mechanics. In application to the creep analysis one should take into account that a general purpose constitutive equation which allows to reflect the whole set of creep and damage processes in structural materials over a wide range of loading and temperature conditions is not available at present. Therefore, a specific constitutive model with selected internal state variables, special types of stress and temperature functions as well as material constants identified from available experimental data should be incorporated into the commercial finite element code by writing a user-defined material subroutine. The examples of manuals for the procedures of user-defined subroutines implementation into the commercial FEM-based software ANSYS and ABAQUS can be found in [3, 22]. The ABAQUS and ANSYS finite element codes are applied to the numerical analysis of creep in structures, e.g. [65, 140, 152, 172, 183, 184].

The standard features of the commercial FEM-based software (e.g., ANSYS and ABAQUS) includes only conventional creep models, refer to [1, 20]. Strain hardening, time hardening, exponential, Graham and Blackburn models, etc. are proposed for the primary creep stage and Garofalo, exponential and Norton models, etc. are proposed for the secondary creep stage. Using standard creep models incorporated into FEM-based software it is impossible to model the tertiary creep stage accompanied with damage accumulation process and fracture, see Fig. 4.1. In order to consider damage processes the user-defined subroutines are developed and implemented. The subroutines serve to utilize constitutive and evolution equations with

4. Creep estimations in structural analysis

damage state variables, see Fig. 4.1. In addition, they allow the postprocessing of damage, i.e. the creation of contour plots visualizing damage distributions.

4.2 Numerical benchmarks for the creep-damage modeling

The concept of benchmarks is widely used in computational engineering mechanics and particularly in creep-damage mechanics. Several benchmark problems based on the different creep constitutive models are presented in [2, 12, 16, 28, 29, 152]. To consider both the creep and the damage processes, a specific constitutive model with selected internal state variables, special types of stress and temperature functions as well as material constants identified from available experimental data should be incorporated into the commercial FE-code by writing a user-defined material subroutine, see e.g. [3, 22]. Thus benchmark problems are needed to verify those developed subroutines. For these problems numerical or analytical reference solutions are usually obtained. To conclude about the fact that the subroutines are correctly coded and implemented, results of finite element computations must be compared with reference solutions of benchmark problems.

4.2.1 Purposes and applications of benchmarks

An important question in the creep analysis is that on reliability of the applied models, numerical methods and obtained results. To assess the reliability of the developed subroutine as well as the accuracy of the results with respect to the mesh density, type of finite element, the time step, and the iteration methods, numerical benchmark problems are required. In [147, 152] the following verification steps are proposed for the reliability assessment:

- *Verification of developed finite element subroutines.* To assess that the subroutines are correctly coded and implemented, results of finite element computations must be compared with reference solutions of benchmark problems. Several benchmark problems have been proposed in [29] based on an in-house finite element code. Below we recall closed form solutions of steady-state creep in elementary structures, well-known in the creep mechanics literature.

4.2. Numerical benchmarks for the creep-damage modeling

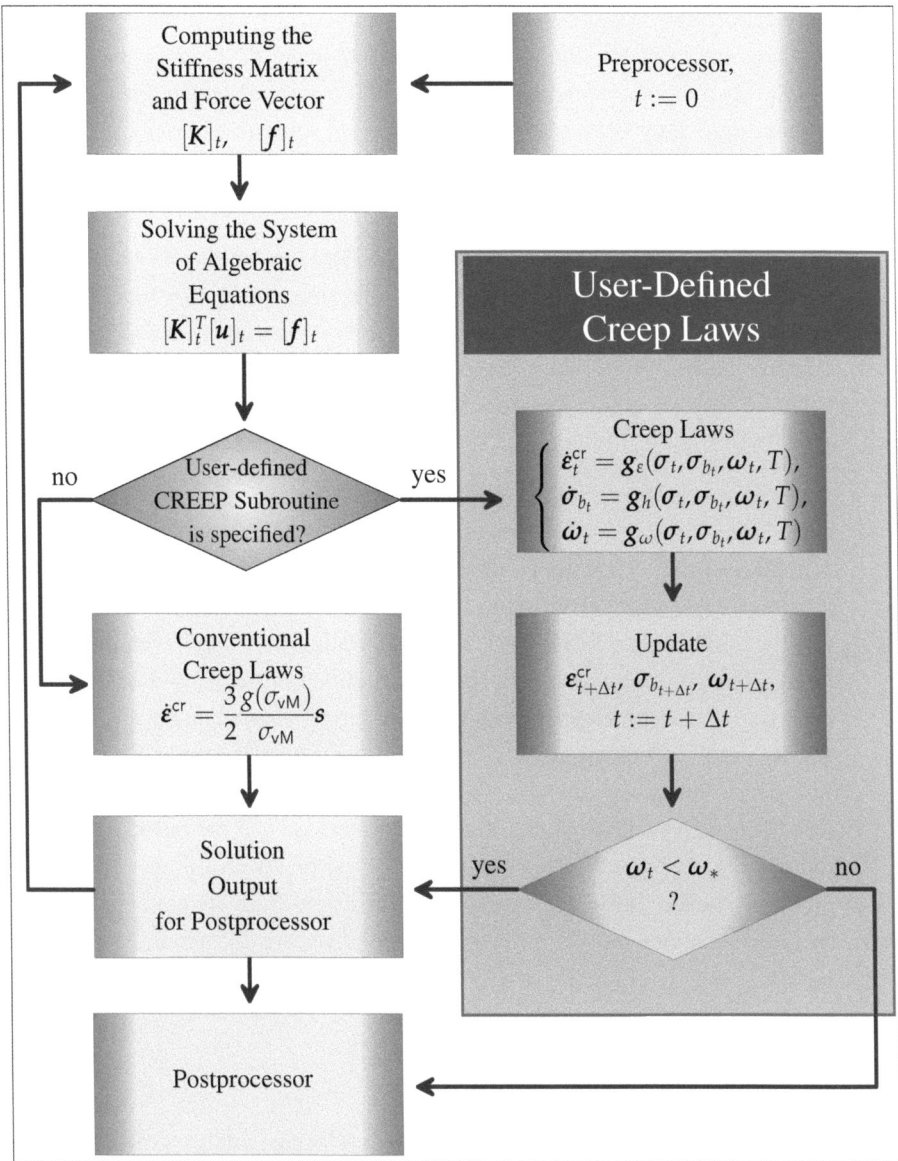

Figure 4.1: Creep analysis procedures in a commercial FEM software with conventional creep laws and with user-defined creep-damage models, after [148].

4. Creep estimations in structural analysis

To extend these solutions to the primary and tertiary creep ranges we apply the Ritz and the time step methods. The advantage of these problems is the possibility to obtain reference solutions without a finite element discretization. Furthermore, they allow to verify finite element subroutines over a wide range of finite element types including beam, shell and solid type elements.

- *Verification of applied numerical methods.* Here the problems of the suitable finite element type, the mesh density, the time step size and the time step control must be analyzed. They are of particular importance in creep damage related simulations. Below these problems are discussed based on numerical tests and by comparison with reference solutions.

- *Verification of constitutive and structural mechanics models.* This step requires creep tests of model structural components and the corresponding numerical analysis by the use of the developed techniques. Examples of recent experimental studies of creep in structures include beams [40, 152], transversely loaded plates [112, 146, 152], thin-walled tubes under internal pressure [115, 117], pressure vessels [65, 67], circumferentially notched bars [86]. Let us note that the experimental data for model structures are usually limited to short-term creep tests. The finite element codes and subroutines are designed to analyze real engineering structures. Therefore long term analysis of several typical structures should be performed and the results should be compared with data collected from engineering practice of power and petrochemical plants. Below the examples of the creep finite element analysis for the typical components of power-generation plants are discussed.

4.2.2 Simply supported beam

To formulate a benchmark problem using the creep-damage analysis of the simply supported aluminium alloy beam loaded by a distributed force q as suggested in [152], it is necessary to compare the results obtained by the Ritz method with those of ABAQUS and ANSYS finite element codes incorporating the Kachanov-Rabotnov

4.2. Numerical benchmarks for the creep-damage modeling

phenomenological creep-damege model [121]:

$$\dot{\boldsymbol{\varepsilon}}^{cr} = \frac{3}{2} \frac{a\,\sigma_{vM}^n}{(1-\omega)^m} \frac{\boldsymbol{s}}{\sigma_{vM}},$$

$$\dot{\omega} = b \frac{[\alpha\,\sigma_T + (1-\alpha)\sigma_{vM}]^k}{(1-\omega)^l} \quad \text{with} \quad \sigma_T = \frac{1}{2}(\sigma_I + |\sigma_I|)$$

(4.2.1)

In this notation $\dot{\boldsymbol{\varepsilon}}^{cr}$ is the creep rate tensor, $\boldsymbol{s}$ is the stress deviator, σ_{vM} is the von Mises equivalent stress, σ_T is the maximum tensile stress, σ_I is the first principal stress, ω is the damage parameter and α represents the weighting factor of a material. $a = 1.35 \cdot 10^{-39}$ [MPa^{-n}/h], $b = 3.029 \cdot 10^{-35}$ [MPa^{-k}/h], $n = 14.37$, $m = 10$, $k = 12.895$, $l = 12.5$ and $\alpha = 0$ are material constants after [116] for the aluminium alloy BS 1472 at 150 ± 0.5°C corresponding to the model (4.2.1). Figure 4.2 shows the good agreement of the time variations of maximum deflection and the normal stress obtained by the Ritz method and the FEM with the exception of the calculations based on the ANSYS finite element code using element type SHELL43, see [21].

4.2.3 Pressurized thick cylinder

We incorporated the following creep constitutive equation (4.2.2) into the ABAQUS finite element code by the means of user-defined material subroutines

$$\dot{\boldsymbol{\varepsilon}}^{cr} = \frac{3}{2} \frac{\dot{\varepsilon}_0}{\sigma_0} \left[1 + \left(\frac{\sigma_{vM}}{\sigma_0} \right)^{n-1} \right] \boldsymbol{s}, \quad \dot{\varepsilon}_0 \equiv a\sigma_0$$

(4.2.2)

with the material constants $\dot{\varepsilon}_0 = 2.5 \cdot 10^{-7}$ 1/h, $\sigma_0 = 100$ MPa, $n = 12$ corresponding to 9Cr1MoVNb steel at 600°C obtained from the creep tests for both the linear and the power law ranges data after [108, 110].

A reference solution of steady-state creep according to Eq. (4.2.2) for a thick cylinder section loaded by internal pressure p is obtained by means of two numerical procedures including the numerical integration and finding the root of a non-linear algebraic equation using Mathcad package (for the details refer to [12]). The results obtained by the presented in [12] approximate method are applied to verify the finite element solution. The steady-state creep problem of the thick cylinder in the power law creep range is the standard benchmark. Thus the geometrical data and the finite element model are assumed as given in the benchmark manual [2].

4. Creep estimations in structural analysis

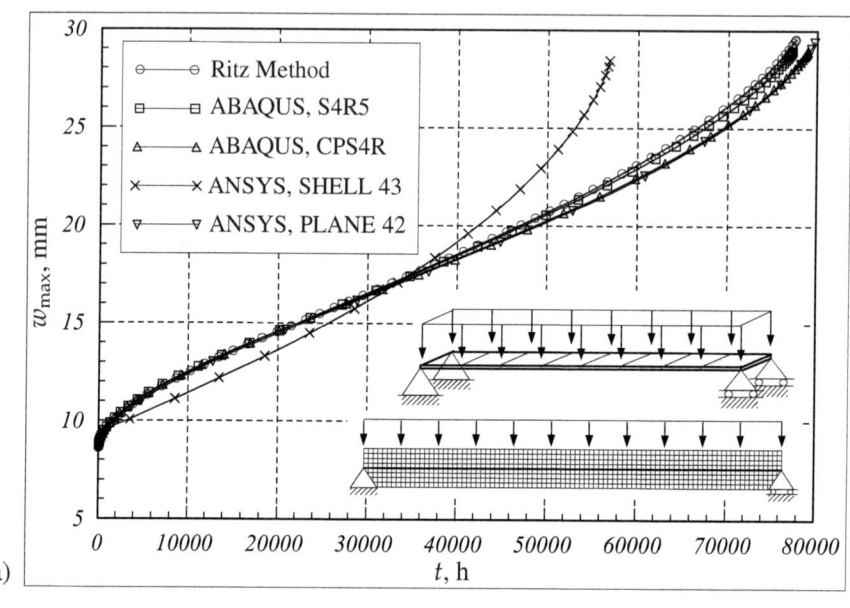

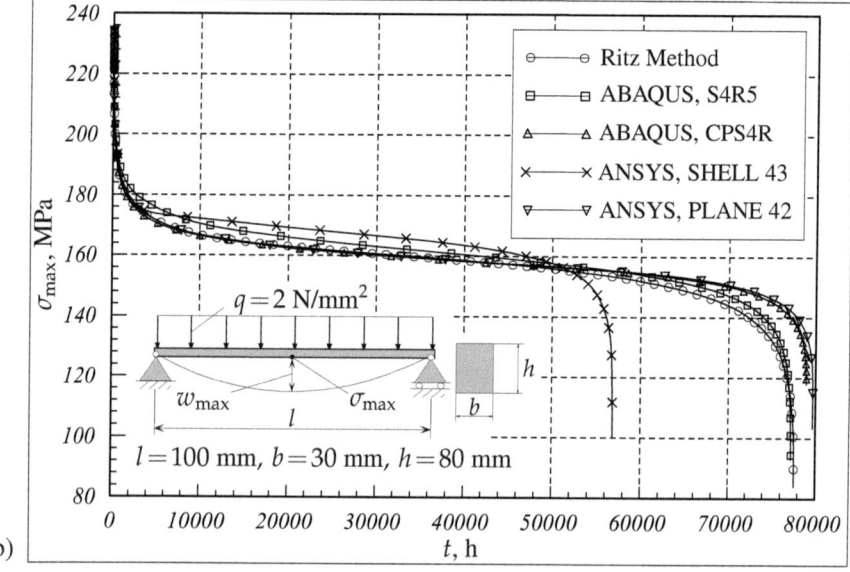

Figure 4.2: Creep-damage analysis results vs. time in the bottom layer of the middle cross-section of a simply supported beam: a) maximum deflection, b) normal stress

Figure 4.3 illustrates a good agreement of the solutions based on the ABAQUS finite code and the approximate numerical solutions for different values of the normalized pressure p in both the linear and the power law ranges.

4.3 Anisotropic creep of a pressurized T-piece pipe weldment

This section is devoted to the practical application of the initially anisotropic model presented in Sect. 2.2.1 and research work [80] to the long-term strength analysis in creep conditions of a T-piece pipe weldment made of the 9Cr-1Mo-V-Nb (ASTM P91) heat-resistant steel. The constitutive model and appropriate creep parameters were based on the experimental creep and rupture data [94] of multi-pass weldmen considering the non-uniformity of the microstructure in the heat affected zone, in the base material and in the weld metal of welding joint. For the purpose of adequate creep behavior numerical modeling of the welding seam, the transversally-isotropic creep-damage model have been incorporated in the commercial FEM-based CAE-software ANSYS. The model takes into account various creep and damage material properties in longitudinal and transversal directions of a welding seam. The performed numerical calculations of the welded structure qualitatively agree with experimental data by MPA Stuttgart [104] and confirm the facts of rupture in corresponding zones of the welding joint.

4.3.1 Formulation of structural model

The typical component of high-temperature power plants and chemical facilities equipment, i.e. pressurized T-piece weldment of two thick-walled pipes with different diameter was chosen as a test example. T-piece pipe weldment is subjected to the internal pressure of 18 MPa and temperature 650°C. Available creep experimental data for this structural component are the results of experimental observations at the T-piece thick-walled pipe weldment creep under the internal pressure 24 MPa at temperature 600°C during 10 000 hours provided by MPA Stuttgart [104]. Geometrical solid model of the welding joint connecting two thick-walled pipes into T-piece weldment was created in CAD-software SolidWorks using the add-on mod-

4. Creep estimations in structural analysis

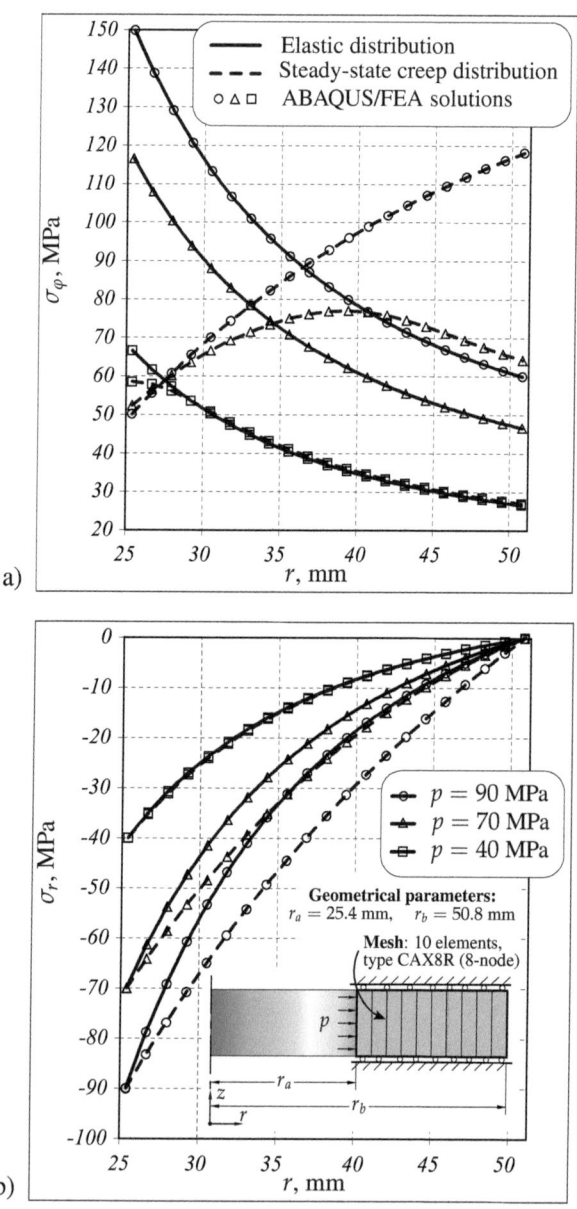

Figure 4.3: Stresses vs. radial coordinate for a pressurized thick cylinder under the steady-state creep conditions: a) hoop stress, b) radial stress

4.3. Anisotropic creep of a pressurized T-piece pipe weldment

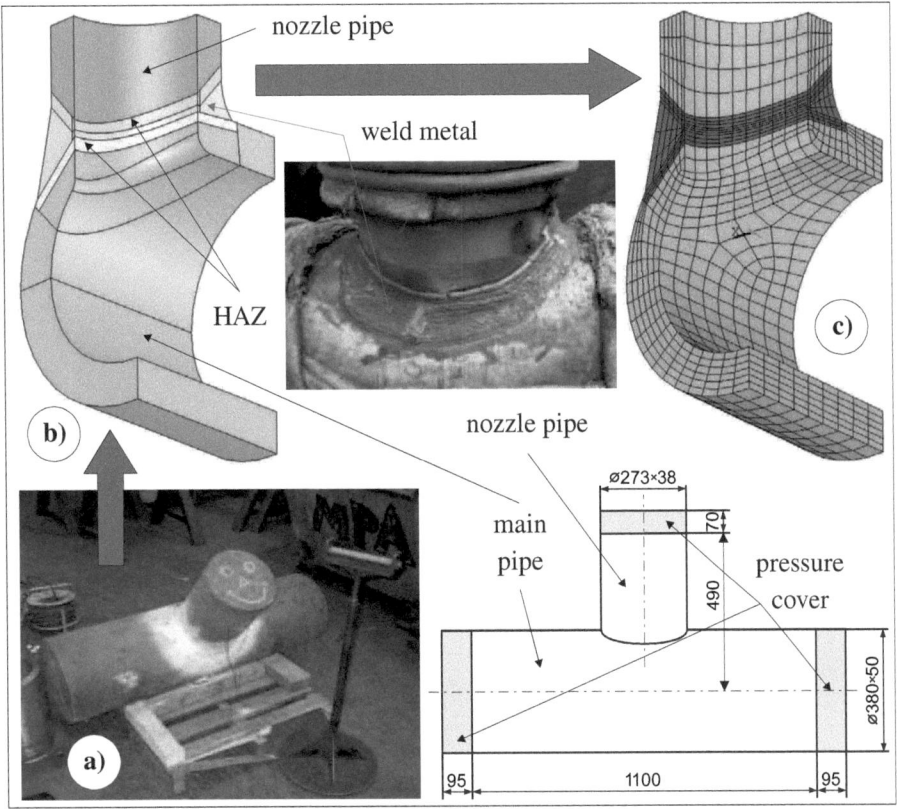

Figure 4.4: Exterior view, geometrical parameters (mm) and structure: a) experimental facility, after [104], b) solid geometry, c) finite-element model.

ule for the design of welded structures [26] basing on the geometrical parameters after MPA Stuttgart [104] and standards GOST [171] and DIN [170], see Fig. 4.4. The finite-element mesh of the T-piece weldment and the series of creep-damage analyses were performed in CAE-software ANSYS.

Basing on previous experience in the creep simulation of welding joints, the weldment is conventionally divided into three zones with different microstructure and creep behaviour character: parent material of pipes, heat-affected zone and welding seam, see Fig. 4.4. But unlike to previous investigations, the initial anisotropy of creep properties in the weld metal of welding seam is considered. Fig. 2.2 illustrates

4. Creep estimations in structural analysis

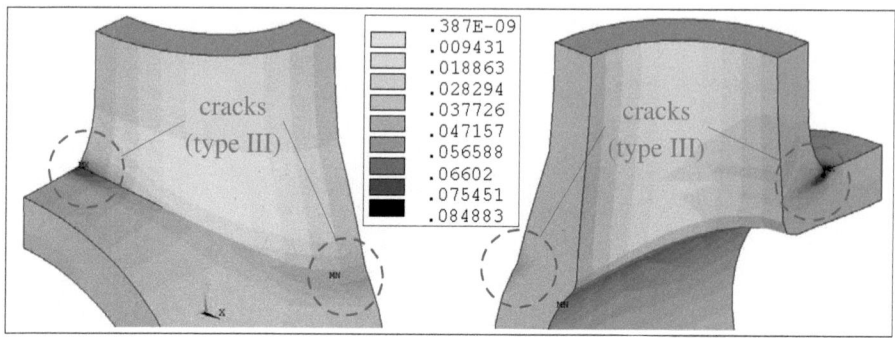

Figure 4.5: Distribution of damage before the rupture: anisotropic parameter ω_1 in the weld metal and isotropic parameter ω in other zones of T-piece pipe weldment.

the creep curves of the 9Cr-1Mo-V-Nb (ASTM P91) steel at 650°C under the tensile stress 100 MPa for three different zones in the welding joint.

4.3.2 Analysis of numerical results

The constitutive creep-damage model (2.2.24) - (2.2.25) with initial anisotropy of creep properties and appropriate creep parameters presented in Sect. 2.2.1 is applied to the long-term strength analysis of the T-piece pipe weldment. In the frames of creep simulations the four life-time assessment with different combinations of creep properties in various structural zones are performed. The first variant of analysis assumes the T-piece pipe weldment being homogenous and consisting of only steel P91 parental material. The second variant of the analysis is based on previous approaches and assumes the weldment consisting of the three materials with different creep behaviour: parent metal of pipe, heat-affected zone and the weld metal properties in the plane of isotropy. And the last two variant of the analysis are similar to the previous ones, however the initial anisotropy of the weld metal creep properties is taken into account.

The creep simulations are performed by the FEM in CAE-software ANSYS till the moment of rupture in each analysis case. And the obtained numerical results demonstrate the specific qualitative tendencies in damage accumulation character as shown below. In the case of homogenous material (parent pipe material) or non-uniform material with isotropic creep properties of all three weldment zones, the

4.3. Anisotropic creep of a pressurized T-piece pipe weldment

failure caused by the damage accumulation occurs only in heat-affected zone, see Fig. 4.5. Such character of failure corresponds to the longitudinal cracks type III due to classification of cracks in weldments presented on Fig. 2.3 and Table 2.1. This type of cracks is proved by the experimental observation presented on Fig. 4.6 and Table 4.1 with circumferential cracks in the heat-affected zone (HAZ) having numbers 1 and 4. In this case the time-to-rupture of the weldment is governed only by the creep properties of the heat-affected zone. The damage accumulation in the isotropic weld metal of welding seam is practically missed, and the value of damage parameter is almost equal to zero. Such a numerical results of creep simulations qualitatively agree with similar investigation in the field of finite-element creep modeling of pressurized welded pipes, see e.g. [90].

Unlike to the previous two analysis cases with isotropic creep properties of materials, the results of the analysis cases considering anisotropic properties of the welding seam show the significant accumulation of two damage parameters ω_1 and ω_2 in the weld metal, see Fig. 4.7. The accumulation character of the damage parameters ω_1 and ω_2 shows the probable initiation of the reach-through longitudinal cracks and transversal cracks. Such character of failure corresponds to cracks types I and II due to classification of cracks in weldments, illustrated on Fig. 2.3 and described in Ta-

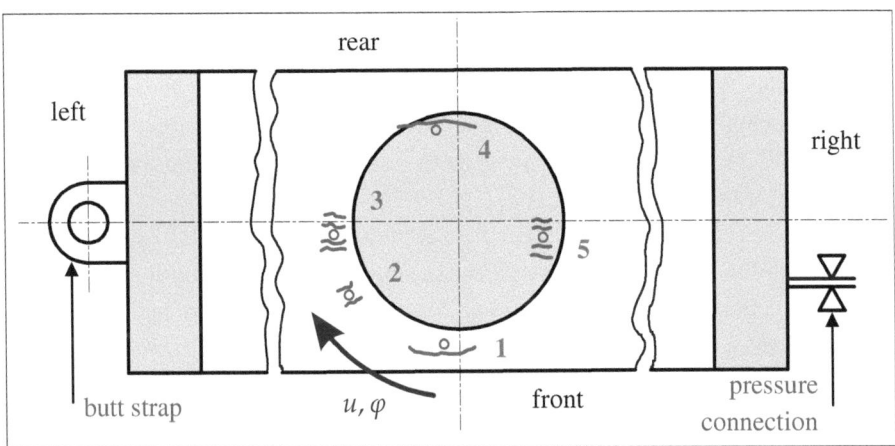

Figure 4.6: Damage in P91 T-piece weldment and welded vessel under internal pressure 24 MPa after 10 000 hours of experimental observations at 650°C, after [104].

4. Creep estimations in structural analysis

Table 4.1: Damage in P91 T-piece weldment and welded vessel under internal pressure 24 MPa after 10 000 hours of experimental observations at 650°C, after [104].

Position	Crack position	Crack direction	Crack center at u, mm	Crack depth, mm
1	HAZ	Circumferential	10	ca. 10
2	Weld metal	Circumferential	150	—
3	Weld metal	T-joint (axial)	220	Through wall
4	HAZ	Circumferential	435	10
5	Weld metal	T-joint (axial)	640	Through wall

ble 2.1. This type of cracks is proved by the experimental observation presented on Fig. 4.6 and Table 4.1 with the reach-through axial cracks in the weld metal having numbers 3 and 5.

The numerical results obtained with the assumption of anisotropic creep properties in the weld metal have more reliable agreement with the creep experimental observations for this t-piece pipe weldment, illustrated on Fig. 4.6 and Table 4.1, and others welded pipe structures [95]. Alongside with the cracks with types III and VI caused by the critical damage accumulation in the heat-affected zone (see Fig. 4.5), the experimental studies [104] report that the reach-through longitudinal cracks with types I and II are initiated by the transversally-isotropic damage parameters accumulation in the weld metal (see Fig. 4.7).

The principal result of the research [80] presented in Sects 2.2.1 and 4.3 is to show the ability of the continuum mechanics approach application to creep simulations of weldments considering non-uniformity of microstructure and anisotropic creep properties of weld metal. Consideration of transversely-isotropic creep properties of weld metal allows to predict the local zones of damage accumulation and probable cracks initiation with better qualitative accuracy. This assumption correlates with the experiments [104] obtained by MPA Stuttgart during the in-service observations for the similar structures. The obtained numerical results demonstrate the necessity of the subsequent creep-rupture experiments for the different weldment

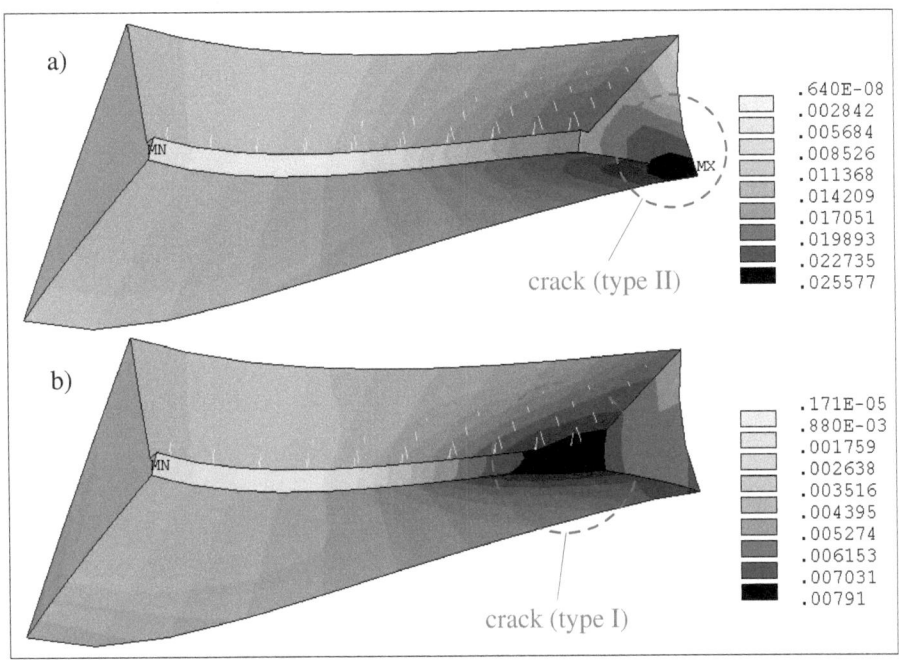

Figure 4.7: Distribution of damage parameters in the weld metal before the rupture: a) along the welding direction ω_1, b) transversely to the welding direction ω_2.

zones (parent material, heat-affected zone and weld metal). The availability of the comprehensive creep experimental data will let to simulate the long-term strength behaviour of weldments more accurately under the in-service loading conditions.

4.4 Creep-damage analysis of power plant components

For the high-temperature components with complex geometry (e.g. pressure piping systems and vessels, rotors and turbine blades, casings of valves and turbines, etc.), where neither analytical nor experimental reference stresses are available, the computer-based finite element analysis (FEA) is used, e.g. [30, 76, 95, 152]. The component is "broken down" into an aggregate of finite elements (FE) with prescribed properties. The computer analysis evaluates the response of the component as a whole and enables the stresses and strains at any given point to be determined. A detailed picture of the results distribution within the component is produced. In its simplest form the FEA may only be applied to analyse the elastic stresses, but the knowledge of the creep properties of the material enables a long-term strength analysis or life-time assessment to be performed.

4.4.1 Previous experience in FEA

The development of computational continuum damage mechanics (CDM) now allows FEA to be performed using physically based constitutive equations to describe the material behavior. This enables the full time dependent behavior of a structure to be modeled, including, by the input of ductility values, the transition from generalized damage to discrete crack growth. Increasing speed and data storage capacity of computer workstations rapidly reduces the time required for such computations. However, the procedure is critically dependent upon the availability of validated multi-axial constitutive equations for deformation and damage over the mechanistic regimes encountered [88, 152].

Modern power-generation plants are required to provide high standards of reliability and availability, which principally depends on the operating conditions, optimal structural design and the applied constructional materials. The reliability and safety of the whole power system depends greatly on the way in which the main power unit components (turbine, boiler and generator) are operated. The operating conditions of those devices are usually very complex and involve many unsteady mechanical and thermal loads which in consequence determine the stress states of the particular components. The durability of the steam turbine should be considered in terms of the durability of its main components, which can be divided into two major groups, mainly the shells (casings, valve housings) and rotating parts (rotors).

The thermal and mechanical loadings within the operation conditions usually lead to degradation of material owing to intensive interactions of many failure modes such as local plasticity, high temperature creep, damage, corrosion, fatigue and cracking. Interactions of main modes of degradation are complex in appropriate numerical modelling and difficult for the numerically based life-time predictions. For instance, creep and damage can be intensified by the process of locally variable high

4.4. Creep-damage analysis of power plant components

thermal stresses during cycles of loading and unloading. Growing demands on safe, reliable and economic operation ask for a sufficient life-time prolongation of the critical power plant elements. A lot of numerical investigations by different laboratories based on CDM and FEA presenting long-term strength analyses and life-time assessments of high-temperature steam-turbine components were published recently, e.g. [38, 39, 98, 158].

In the framework of this dissertation several FEM-based life-time assessments of high-temperature power plant components were performed using the formulated in Sect. 2 constitutive creep-damage models:

- The numerical investigations of long-term strength behaviour started with the application of conventional Kachanov-Rabotnov-Hayhurst creep-damage model [40, 88, 152] to isothermal creep-damage simulation [77]. As the component to be analysed the casing of a steam turbine reheat control valve was chosen. Such a component has to control the steam flow into an intermediate pressure steam turbine. The investigation has proved the high efficiency of FEM-based life-time assessments with the application of conventional CDM-based creep models.

- Within the frames of the research work [79] two conventional creep-damage models were applied to the isothermal analysis of the mechanical behaviour of a steam turbine rotor under in-service conditions. These models were the isotropic Kachanov-Rabotnov-Hayhurst model [40, 88, 152] and Murakami-Ohno [141, 142] model with damage induced anisotropy presented in Sect. 2.2.2. Numerical solutions of the initial-boundary value problems have been obtained by FEM using solid axisymmetrical type finite elements. For the purpose of long-term strength analysis both isotropic and anisotropic creep-damage models have been implemented in the commercial FE-code ANSYS. Obtained simulation results for a steam turbine rotor show the significant sensitivity of life-time assessment to the type of material model.

- Thereafter, the conventional Kachanov-Rabotnov-Hayhurst creep-damage model [40, 88, 152] was extended to the case of variable temperature and strain hardening consideration, as presented in Sect. 2.1. A technique for the identification of material creep constants based on the available family of experimental creep curves (see Appendix A) was presented in [114, 129]. The resulting non-isothermal model with the appropriate creep material parameters [113] were incorporated into the FE-code of the CAE-software ABAQUS. It was applied to the long-term strength analysis of a pressurized thick-walled pipe exposed to the non-uniform heating for the verification purposes and for the illustration of basic features of creep-damage character, refer to [129]. Than the model was applied to the non-isothermal life-time assessments under in-service loading conditions of such power plant components, as the casing of a steam turbine control-stop valve [114], the casing of a steam turbine bypass valve [130], housing of a high-pressure gas turbine [131], and the casing of a steam turbine quick-stop valve [132]. The obtained numerical results confirmed the influence on non-uniform temperature field on the final stress and strain fields redistribution before the failure. However, the time-to-rupture values under the in-service loading conditions were overestimated, because the applied model is suitable only for the narrow range of the high stresses.

4.4.2 Steam turbine quick-stop valve

Finally, we must discuss several results of long-term analysis obtained by application of the new creep-damage constitutive model developed in Chapt. 3 to creep behaviour simulation of a power plant component. As the component to be analysed the casing of a steam turbine quick-stop valve [53–55], illustrated on Fig. 4.8, was chosen.

Application

If the steam turbine is tripped it is essential to have a fast and reliable quick stop function for protection of the turbine. The steam turbine quick stop valve (VQS) is designed for this purpose [54]. A mechanical spring is used for the emergency shut-off. The closing time is typically less than 0.2 s for full stroke. On request shorter closing time is available. The emergency quick close function achieves maximum protection for the steam turbine. The valves are equipped with on-line exercise capability that demonstrates freedom of movement of tripping components without affecting steam flow. The VQS valve is used together with the turbine flow control

4.4. Creep-damage analysis of power plant components

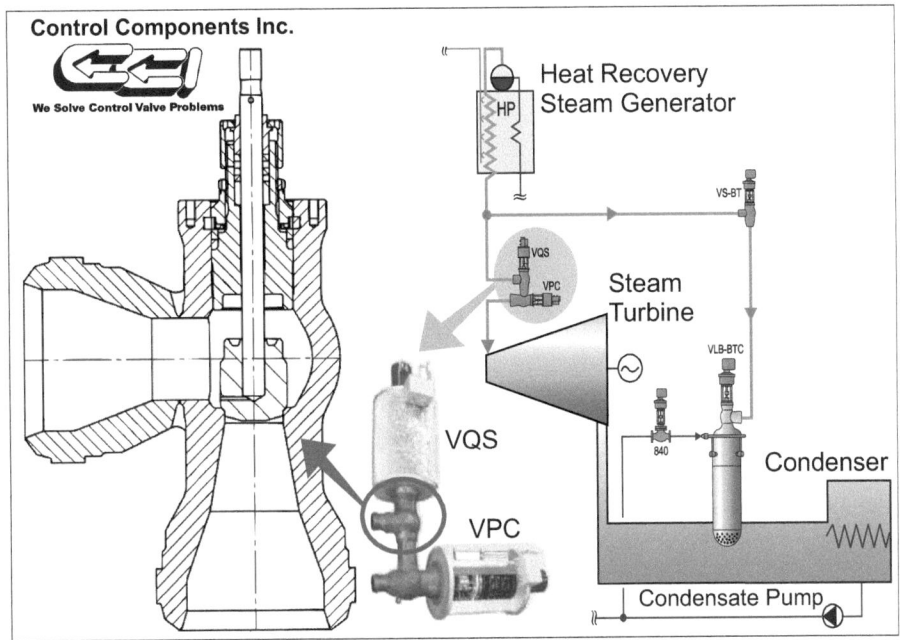

Figure 4.8: Typical installation of the steam turbine quick stop valve in a power station, after [54, 55].

valve (VPC), as illustrated in Fig. 4.8. The VPC valve is designed for regulating the steam flow from start-up to full load. As an option the VPC valve can be equipped with a quick closing function. The low total pressure drop over the valves improves the energy efficiency of the installation, thus improving the power production. Each valve is independently calculated and designed according to the relevant operational conditions.

Design features

The VQS and VPC valves design is of angle type and is based on the well proven design of stop valve type VS. As far as, the drawings of the VQS valve casing were not accessible from opened sources, it was decided to use the drawing of VS valve casing available from [55] and illustrated on Fig. 4.8 for the modeling of 3D solid geometry in the CAD-software SolidWorks. The VS valve casing [55], fully machined of forged CrMo low alloy steel or carbon steel, including X10CrMoVNb91 (F91) steel, is designed to minimize material stresses as well as to fit the requirements of the piping system with regard to material, pressure class and piping connections [53]. An even material distribution is essential to minimize the material stresses. By using the homogenous forged material, an accurate and controlled wall thickness, i.e. a smooth surface, is achieved. For severe operating conditions with large temperature variations, a continuous preheating of the valve inlet side is recommended. The VS valve is designed to be operated by an actuator, and depending on the actuating force and project preference, any type of actuator can be selected: pneumatic, electrohydraulic or electromechanical. The unbalanced configuration of VS valve casing with tight design providing leakage tightness according to ANSI B16.104 Class V was selected, see Fig. 4.8. The maximum capacity of the VQS valve is 7500 Kv or 8660 Cv, the maximum inlet pressure class corresponds to DIN-PN 640 (ANSI# 4500), the maximum outlet pressure class corresponds to DIN-PN 250 (ANSI# 1500). Flow path in angle body of the VQS valve means low pressure drop due to pressure recovery in the outlet cone, refer to [53].

Review of FEA results

The 3D-solid geometry of the valve casing designed in the CAD-software SolidWorks is transferred to the FEM-based CAE-software ABAQUS and meshed with 7740 solid 8-noded finite elements (type 3CD8R, refer to [1]) providing totally 10085 nodes, as illustrated on Fig. 4.9. The analysed main part of the valve casing is simplified to symmetrical geometry with applied symmetry boundary conditions on the cross-section surface. The main load on the valve casing is the internal pressure 20 MPa which is considered as constant over time and normal for operating conditions due to the service applications. As far as the geometry of a valve actuator is not taken into account, its effect is replaced with balance pressure p_{noz} applied to the outer surface of nozzle with opposite sign, as illustrated on Fig. 4.9. It is calculated as follows

$$p_{noz} = \frac{p_{int} \cdot A_{act}}{A_{noz}} = 8.67 \text{ (MPa)}, \tag{4.4.3}$$

where $p_{int} = 20$ MPa is the internal pressure in the valve, $A_{act} = 1.24 \cdot 10^4$ mm² is the area of the actuator cross-section and $A_{noz} = 2.85 \cdot 10^4$ mm² is the area of the actuator cross-section.

4.4. Creep-damage analysis of power plant components

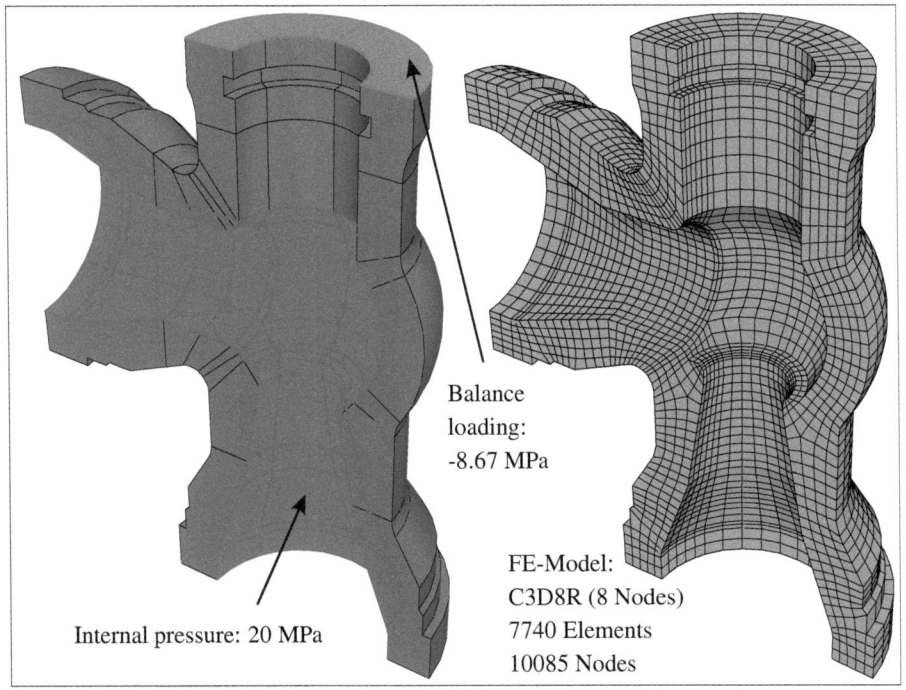

Figure 4.9: ABAQUS model of VQS: geometry, loadings and FE-mesh.

Additional damage due to fatigue caused by the transient operation modes of a steam turbine, e.g. during start-up and shut down, is not considered here. The modern experience of steam turbines manufacturers [101] shows the 9-12%Cr advanced heat-resistant steels are used for valve casings or turbine housings manufactured by casing or forging processes. Thus, the material of the analyzed valve casing is assumed to be the 9Cr-1Mo-V-Nb (ASTM P91) heat-resistant steel. As the valve is considered to operate at temperature 600°C, the values of Young's modulus $E = 1.12 \cdot 10^5$ MPa, Poisson's ration $\mu = 0.3$ and thermal expansion coefficient $\alpha = 1.26 \cdot 10^{-5}$ K^{-1} are taken from [65]. Finally, for the transient creep process simulation, the isothermal form of the creep constitutive model developed in Chapt. 3 is applied to predict the creep damage in the valve casing. The creep constitutive equation (3.4.45) with the strain-hardening function (3.4.46) and the damage evolution equations (3.4.47) with the time-to-rupture function (3.4.48) including all the

4. Creep estimations in structural analysis

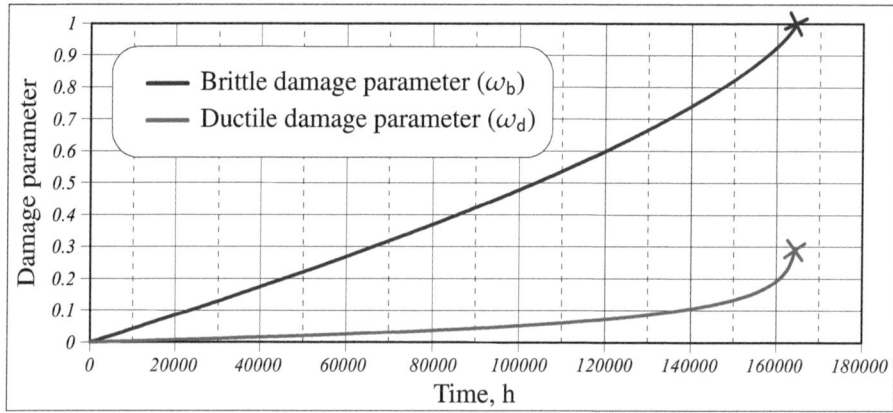

Figure 4.10: Damage accumulation character of brittle (ω_b) and ductile (ω_d) parameters.

corresponding creep materials parameters summarized in Sect. 3.4 are incorporated in FE-code of CAE-software ABAQUS by the means of user-defined creep subroutine. The proposed model is able to reflect the basic features of stress redistribution in the structural component. Furthermore, it allows us to predict the locations for the maximum creep damage and the life-time of the structural component till failure.

The creep-damage behavior simulation of the valve casing with the previously mentioned initial-boundary conditions and material properties has predicted the failure of the component in $t^* = 164000$ hours $= 18.7$ years. The obtained FEA results show the critical damage accumulation in two locations. The first location is situated on the outer surface of the valve casing as illustrated on Fig. 4.12 and is caused by the brittle damage parameter ω_b critical concentration. In this possible place of brittle rupture initiation the damage accumulation is dominantly governed by the maximum tensile stress $\sigma_{max\ t}$, as shown on Fig. 4.12. The second location is situated on the inner surface of the valve casing as illustrated on Fig. 4.13 and is caused by the ductile damage parameter ω_d critical concentration. In this possible place of ductile rupture initiation the damage accumulation is dominantly governed by the von Mises effective stress σ_{vM}, as shown on Fig. 4.13. As far as the character of the both damage parameters (ω_b and ω_d) evolution is found out to be equal in the both locations (see Fig. 4.10), the decision about the type of rupture is done basing on the stress param-

4.4. Creep-damage analysis of power plant components

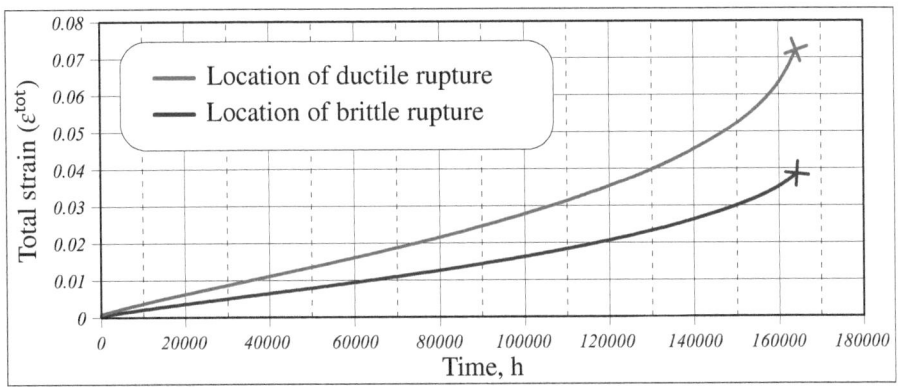

Figure 4.11: Accumulation characters of maximum principal total strain (ε^{tot}) in the locations of rupture.

eters ($\sigma_{max\,t}$ and σ_{vM}) redistribution character. The dominant stress parameter ($\sigma_{max\,t}$ or σ_{vM}) defines the type of rupture (brittle or ductile). Additionally, this assumption is proved by the different evolution character of the maximum principal total strain ε^{tot} in different rupture locations, illustrated on Fig. 4.11. The comparison of the creep curves shows, that the ductile rupture location has accumulated almost two times more creep strain than the brittle rupture location. Due to Fig. 4.11, the ductile rupture location has more prevalent tertiary creep stage of the creep curve, but in the both locations the rupture occurs in the same time t^*.

These results should be investigated further in order to find out how to come closer to reality. Such clarification could give important input for future improvements in high temperature component design. In this context, firstly the parameter identification should be optimized for the relevant loading conditions, whereby also multi-axial experiments should be used. Moreover, it is very important to compare and review the numerical damage predictions with respect to experimental findings of uni-axial and multi-axial load cases.

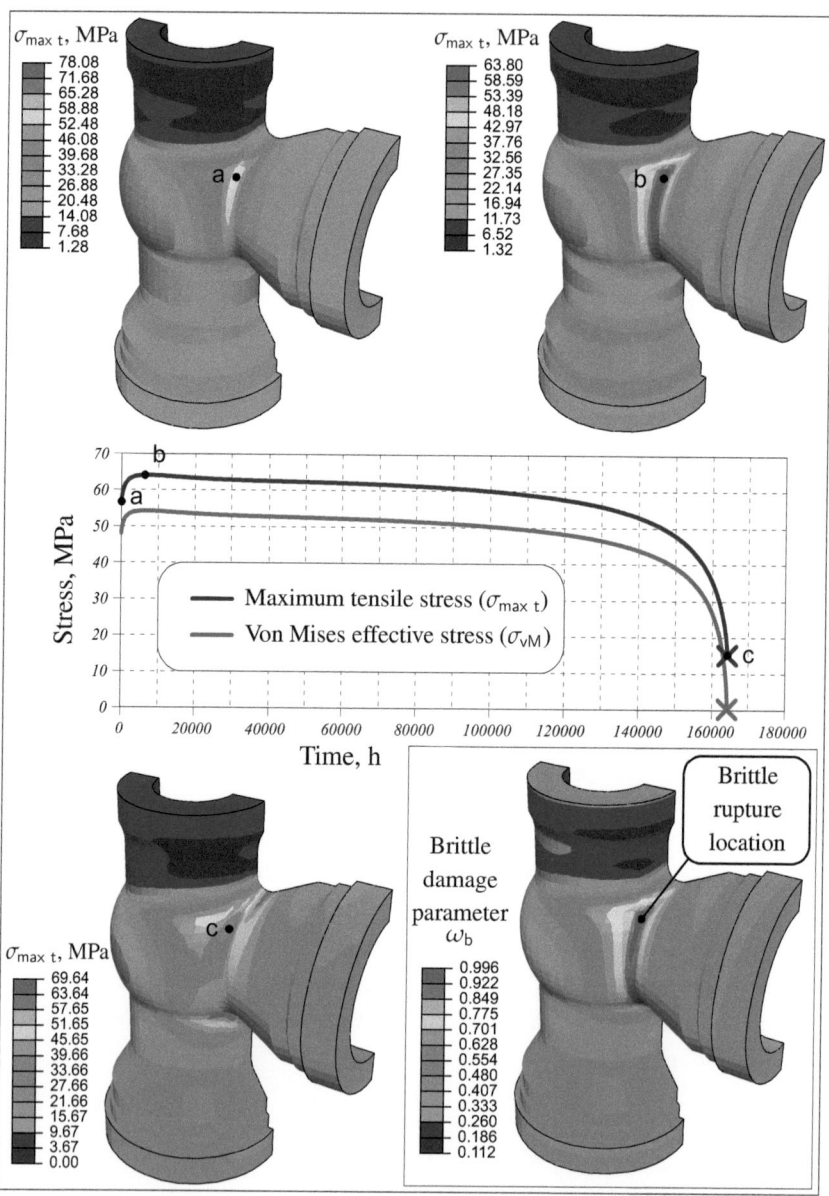

Figure 4.12: Redistribution of the maximum tensile stress $\sigma_{\text{max t}}$ in the location of brittle rupture.

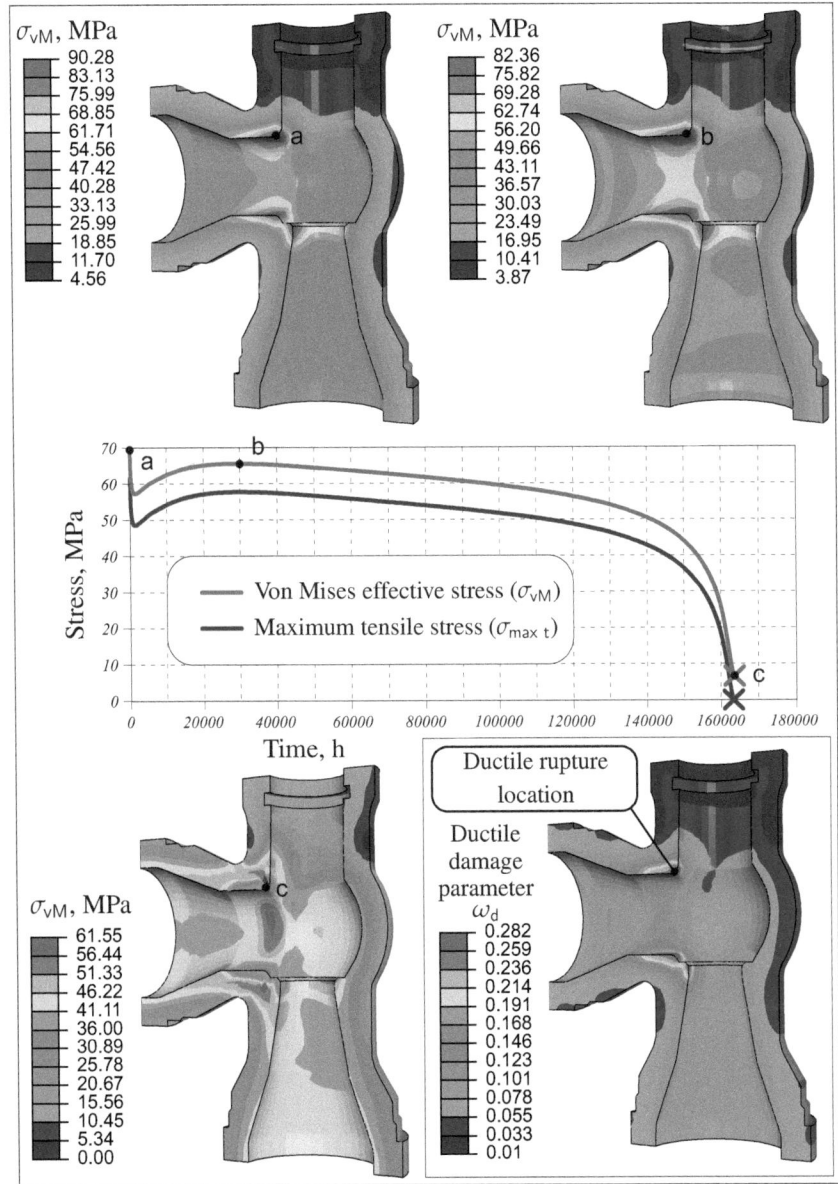

Figure 4.13: Redistribution of the von Mises effective stress σ_{vM} in the location of ductile rupture.

Chapter 5

Conclusions and Outlook

A new appropriate approach to the phenomenological modeling of creep and damage behaviour based on the well-known concepts of continuum damage mechanics and creep mechanics is developed. In the framework of the thesis a comprehensive non-isothermal creep-damage model for a wide stress range is proposed in Chapt. 3. It is based on the available creep and rupture experimental studies and microstructural observations for advanced heat resistant steels. The proposed approach takes into account the following features, which are important for the comprehensive creep and damage modeling for structural analysis of industrial components:

- The uniaxial form of the proposed creep-damage model describing primary, secondary and tertiary creep stages is formulated. The model consists of the constitutive equation for the creep strain rate governing steady-state creep behaviour and two damage evolution equations for accumulation rates of brittle and ductile damage parameters governing tertiary creep behaviour and rupture. To take into account the primary creep behavior and stress relaxation effects a strain hardening function is utilized in the constitutive equation.

- The creep constitutive equation presented in Sect. 3.1 shows the stress range dependent behavior presenting the power-law to linear creep mechanism transition with a decreasing stress. To take into account the primary creep behavior a strain hardening function is utilized in Sect. 3.2. The constitutive equation in the form of serious connection of linear and power-law components extended with strain hardening function well describes the creep strain rates for a wide stress range and stress relaxation process under the loading values relevant to

5. Conclusions and Outlook

in-service conditions of industrial applications.

- To characterize creep-rupture behavior the constitutive equation is generalized by introduction of two damage internal state variables and appropriate evolution equations in Sect. 3.3. The description of long-term strength behavior is based on the assumption of ductile to brittle damage character transition with the decrease of stress. Two damage parameters show different ductile and brittle damage accumulation characters based on the Kachanov-Rabotnov concept, but the similar time-to-rupture dependence.

- Since the creep and the damage effect are heat-activated processes, the creep constitutive and damage evolution equations are extended with the temperature dependence using the Arrhenius-type functions. Such a temperature dependence presented in Sects 3.1 and 3.3 is found applicable only for a quite narrow range of high temperatures.

- The set of creep material parameters for the 9Cr-1Mo-V-Nb (ASTM P91) heat-resistant steel valid in the stress (0 – 300 MPa) and temperature (550 – 650°C) ranges relevant to in-service loading conditions is identified in Sect. 3.3. The identification procedure is based on manual fitting of the available experimental data presented by the minimum creep strain rate, time-to-rupture and creep-rupture strain vs. stress dependencies.

- The unified multi-axial form of the creep-damage model is presented in Sect. 3.4. To analyze the failure mechanisms under multi-axial stress states the isochronous rupture loci and time-to-rupture surface are presented. They illustrate that the proposed failure criterion based on the dependence of time-to-rupture on stress include both the maximum tensile stress and the von Mises effective stress. But the measures of influence of the stress parameters are dependent on the level of stress with ductile to brittle failure transition.

- Finally in Sect. 4.4 an example of numerical long-term strength analysis for a typical power plant component is presented. The obtained FEA results show that the developed approach is capable to reproduce basic features of creep and damage processes in engineering structures. And the assumptions of power-law to linear creep behaviour and ductile to brittle damage mode transitions are really important for an adequate life-time assessments and local damaged zones predictions under in-service loading conditions.

The proposed in this work creep-damage model is not yet well verified and optimized. The form of strain hardening function for the description of primary creep stage proposed in Sect. 3.2 is still under question. Because the primary stages of the creep curves do not fit the experimental creep curves equally well under low and high levels of stress. Therefore, for the purpose of testing and improvement the following future studies are planed:

- The strain hardening function proposed in Sect. 3.2 probably need to be replaced with other more suitable concepts for the purpose of better description of primary creep stage. Possible variants are the modification of the strain hardening function with stress-dependent primary creep parameters or the application of kinematic hardening concepts, see e.g. [152]. The common approach used in the kinematic hardening concept is the introduction of additional internal state variables such as back stress [134, 135] and hardening [46, 167] and appropriate evolution equations (the so-called hardening rules).

- The verification of the creep-damage model presented in Chapt. 3 should be performed using solutions of benchmark problems specified for creep and damage mechanics. The main problem while the model development was the lack of creep and rupture experimental data under the low level of stress. Thus, the accuracy of the creep-damage model predictions can be increased by a new creep material parameters identified using of experiments obtained under low and moderate stresses and higher temperature range. The independent creep-rupture tests under non-proportional loading (e.g. combined tension and torsion) in a wide stress range are necessary to verify the new failure criterion.

- The algorithm and the procedure for the automatical or semi-automatical identification of the creep materials parameters applicable for any of the similar advanced heat-resistant steels have to be developed. An application or a stand-alone software for the identification should automatically fit the input experimental data for the series of stresses and temperatures and must output the

5. Conclusions and Outlook

values of creep material parameters. The identification procedure can be additionally visualized and extended with queries to final user to define an identification options and to optimize the values of creep parameters.

Chapter A

Identification procedure of creep material parameters

A.1 Secondary creep stage

The steady-state creep region of every creep curve from experimental data set $(\varepsilon_d, t_d), (\varepsilon_{d+1}, t_{d+1}), \ldots, (\varepsilon_q, t_q)$, where $d \geq 2$ and $q \leq f$ (f is a number of experimental measurements), is approximated by least-squares regression method [200], using linear function

$$\varepsilon = \dot{\varepsilon}^{cr}_{min} \, t, \tag{A.1.1}$$

where the minimum creep strain rate $\dot{\varepsilon}^{cr}_{min}$ is the slope to be estimated from linear function (A.1.1) in the following form:

$$\dot{\varepsilon}^{cr}_{min} = \frac{\left[(q-d)\sum_{i=d}^{q} t_i \varepsilon_i - \left(\sum_{i=d}^{q} t_i\right)\left(\sum_{i=d}^{q} \varepsilon_i\right)\right]}{\left[(q-d)\sum_{i=d}^{q} t_i^2 - \left(\sum_{i=d}^{q} t_i\right)^2\right]}. \tag{A.1.2}$$

The material constants A and n for the relationship between creep strain rate and stress $\dot{\varepsilon}^{cr} = A\,\sigma^n$ can be determined from steady-state creep. In double logarithmic coordinates the creep strain rate and stress theoretically must be connected by close to linear function, which can be approximated by least-squares regression method [200], using linear function

$$\lg \dot{\varepsilon}^{cr}_{min} = \lg A + n \lg \sigma, \tag{A.1.3}$$

A. Identification procedure of creep material parameters

where the creep exponent n is the slope of function (A.1.3), linear in double logarithmic scale.

Specifying $\dot{\varepsilon}^{cr}_{\min 1}, \dot{\varepsilon}^{cr}_{\min 2}, \ldots, \dot{\varepsilon}^{cr}_{\min \xi}$ as minimum creep rates at the fixed stresses $\sigma_1, \sigma_2, \ldots, \sigma_\xi$ and fixed temperatures $T_1, T_2, \ldots, T_\varphi$, where ξ is a number of experimental stresses and φ is a number of temperatures corresponding to experimental creep curves with secondary creep stage, constant n and the set of temperature dependent constants A_j ($j = 1, 2, \ldots, \varphi$) can be estimated from following relations, produced by least-squares regression method [200], using approximation function (A.1.3):

$$n = \frac{\xi \sum_{i=1}^{\xi} (\lg \dot{\varepsilon}^{cr}_{\min i} \; \lg \sigma_i) - \left(\sum_{i=1}^{\xi} \lg \dot{\varepsilon}^{cr}_{\min i}\right) \left(\sum_{i=1}^{\xi} \lg \sigma_i\right)}{\xi \sum_{i=1}^{\xi} \lg \sigma_i^2 - \left(\sum_{i=1}^{\xi} \lg \sigma_i\right)^2}, \quad (A.1.4)$$

$$\lg A_j = \frac{\left(\sum_{i=1}^{\xi} \lg \dot{\varepsilon}^{cr}_{\min i}\right) \left(\sum_{i=1}^{\xi} \lg \sigma_i^2\right) - \left(\sum_{i=1}^{\xi} (\lg \dot{\varepsilon}^{cr}_{\min i} \; \lg \sigma_i)\right) \left(\sum_{i=1}^{\xi} \lg \sigma_i\right)}{\xi \sum_{i=1}^{\xi} \lg \sigma_i^2 - \left(\sum_{i=1}^{\xi} \lg \sigma_i\right)^2}. \quad (A.1.5)$$

The data array (A_j, T_j), where $j = 1, 2, \ldots, \varphi$, representing temperature dependence of secondary creep can be quite accurately approximated by the least-squares regression method [200], using Arrhenius law function

$$A(T) = a \, \exp\left(-\frac{h}{T}\right), \quad (A.1.6)$$

which is linear in half-logarithmic scale, and which contains the creep material constants a and h, defined by notations

$$a = \frac{A(T_1)}{\exp\left(-\dfrac{h}{T_1}\right)} = \frac{A(T_2)}{\exp\left(-\dfrac{h}{T_2}\right)} \quad \text{and} \quad h = \frac{T_1 T_2 \ln\left[\dfrac{A(T_1)}{A(T_2)}\right]}{T_1 - T_2}. \quad (A.1.7)$$

A.2 Tertiary creep stage

Material constants l_j define the shape of creep curve on the tertiary creep stage and governs the value of critical creep strain $(\varepsilon_1^*, \varepsilon_2^*, \ldots, \varepsilon_\psi^*)$ corresponding to rupture time $(t_1^*, t_2^*, \ldots, t_\psi^*)$ and stresses $(\sigma_1, \sigma_2, \ldots, \sigma_\psi)$ for the number of fixed temperatures $T_1, T_2, \ldots, T_\varphi$:

$$l_i(T_j) = n - 1 + \frac{n}{\left[\frac{\varepsilon_i^*}{A \exp(-h/T_j) \, (\sigma_i^*)^n \, t_i^*}\right] - 1} \quad (A.2.8)$$

$$\text{for} \quad (j = 1, 2, \ldots, \varphi) \quad \text{and} \quad (i = 1, 2, \ldots, \psi),$$

where ψ is a number of experimental stress values corresponding to experimental creep curves with tertiary creep stage and φ is a number of temperature values.

The creep curves with an evident tertiary creep stage $(\sigma_1, \sigma_2, \ldots, \sigma_\psi)$ are selected from the complete experimental stress range. The value of l_j $(j = 1, 2, \ldots, \varphi)$ for every temperature value T_j is defined as an arithmetic mean value of the array $l_i(T_j)$ $(i = 1, 2, \ldots, \psi)$. For all temperatures $T_1, T_2, \ldots, T_\varphi$ the value of creep constant l is defined as a simple average value of the set l_j $(j = 1, 2, \ldots, \varphi)$.

The long-term strength curves corresponding to the number of fixed temperatures $T_1, T_2, \ldots, T_\varphi$, i.e. failure time $(t_1^*, t_2^*, \ldots, t_\psi^*)$ vs. applied stress $(\sigma_1, \sigma_2, \ldots, \sigma_\psi)$ relations, can be approximated by the least-squares regression method [200], using the following function, linear in a double logarithmic scale:

$$\lg t^* = -\lg\left[(l+1)\,B\right] - m\,\lg\sigma, \quad (A.2.9)$$

where ψ is a number of experimental stress values corresponding to the available experimental creep curves with tertiary creep stage. The constant m, which is not dependent on the temperature, characterized the slope of the function (A.2.9), linear in a double logarithmic scale.

The constant m and the set of temperature dependent constants B_j $(j = 1, 2, \ldots, \varphi)$ can be estimated from following relations, produced by the least-squares regression

A. Identification procedure of creep material parameters

method [200], using approximation function (A.2.9):

$$m = -\frac{\psi \sum_{i=1}^{\psi} (\lg t_i^* \, \lg \sigma_i) - \left(\sum_{i=1}^{\psi} \lg t_i^*\right) \left(\sum_{i=1}^{\psi} \lg \sigma_i\right)}{\psi \sum_{i=1}^{\psi} \lg \sigma_i^2 - \left(\sum_{i=1}^{\psi} \lg \sigma_i\right)^2}, \qquad (A.2.10)$$

$$\lg\left[(l+1)\, B_j\right] = \\ -\left[\frac{\left(\sum_{i=1}^{\psi} \lg t_i^*\right)\left(\sum_{i=1}^{\psi} \lg \sigma_i^2\right) - \left[\sum_{i=1}^{\psi}(\lg t_i^* \, \lg \sigma_i)\right]\left(\sum_{i=1}^{\psi} \lg \sigma_i\right)}{\psi \sum_{i=1}^{\psi} \lg \sigma_i^2 - \left(\sum_{i=1}^{\psi} \lg \sigma_i\right)^2}\right]. \qquad (A.2.11)$$

The data array (B_j, T_j), where $j = 1, 2, ..., \varphi$, representing temperature dependence of tertiary creep can be quite accurately approximated by the least-squares regression method [200], using Arrhenius law function

$$B(T) = b \, \exp\left(-\frac{p}{T}\right), \qquad (A.2.12)$$

which is linear in half-logarithmic scale, and contains creep material constants b and p, defined by the following notations:

$$b = \frac{B(T_1)}{\exp\left(-\dfrac{p}{T_1}\right)} = \frac{B(T_2)}{\exp\left(-\dfrac{p}{T_2}\right)} \quad \text{and} \quad p = \frac{T_1 T_2 \ln\left[\dfrac{B(T_1)}{B(T_2)}\right]}{T_1 - T_2}. \qquad (A.2.13)$$

A.3 Primary creep stage

The analytical formulation of the primary creep strain can be provided, if in Eq. (2.1.8) the influence of secondary and tertiary creep stages are neglected as follows:

$$\varepsilon = k \ln\left[(1+C) \exp\left(\frac{\theta}{k}\right) - C\right] \quad \text{with} \quad \theta = a \exp\left(-\frac{h}{T}\right) \sigma^n t, \qquad (A.3.14)$$

where C and k are primary creep material constants to be defined.

A.3. Primary creep stage

For all the number of fixed temperatures $T_1, T_2, \ldots, T_\varphi$ (φ is the number of temperature values) the creep curves with well marked primary creep stage are selected. Then for all selected creep curves the maximum creep strain values at the end of primary creep stage are obtained for all fixed temperatures: $(\varepsilon_{\max 1}^{T_1}, \varepsilon_{\max 2}^{T_1}, \ldots, \varepsilon_{\max \phi}^{T_1})$, $(\varepsilon_{\max 1}^{T_2}, \varepsilon_{\max 2}^{T_2}, \ldots, \varepsilon_{\max \phi}^{T_2}), \ldots, (\varepsilon_{\max 1}^{T_\varphi}, \varepsilon_{\max 2}^{T_\varphi}, \ldots, \varepsilon_{\max \phi}^{T_\varphi})$. The next step assumes the extraction of φ creep curves with close or equal maximum creep strain values, one for every temperature value. For example, the following array is obtained after extraction: $\varepsilon_{\max 1}^{T_\varphi}, \varepsilon_{\max 2}^{T_2}, \ldots, \varepsilon_{\max \phi}^{T_1}$. Creep curves corresponding to the array $(\varepsilon_{\max 1}^{T_\varphi}, \varepsilon_{\max 2}^{T_2}, \ldots, \varepsilon_{\max \phi}^{T_1})$ are approximated by function (A.3.14) using least-squares regression method [200] for the purpose of estimation of primary creep constants C_δ and k_δ ($\delta = 1, 2, \ldots, \varphi$) for every temperature value $T_1, T_2, \ldots, T_\varphi$.

The final values of C and k are obtained as simple average values for arrays C_δ and k_δ ($\delta = 1, 2, \ldots, \varphi$) as follows

$$C = \frac{C_1 + C_2 + \ldots + C_\varphi}{\varphi} \quad \text{and} \quad k = \frac{k_1 + k_2 + \ldots + k_\varphi}{\varphi}. \tag{A.3.15}$$

Bibliography

[1] ABAQUS, Inc., Dassault Systèmes, *ABAQUS Analysis User's Manual*, Version 6.7 ed., 2007.

[2] ABAQUS, Inc., Dassault Systèmes, *ABAQUS Benchmarks Manual*, Version 6.7 ed., 2007.

[3] ABAQUS, Inc., Dassault Systèmes, *ABAQUS User Subroutines Reference Manual*, Version 6.7 ed., 2007.

[4] ABE, F., "Creep rates and strengthening mechanisms in tungsten-strengthened 9Cr steels," *Materials Science and Engineering*, vol. A319-A321, pp. 770 – 773, 2001.

[5] ABE, F., "Bainitic and martensitic creep-resistant steels," *Current Opinion in Solid State and Materials Science*, vol. 8, no. 3-4, pp. 305 – 311, 2004.

[6] ABE, F. and TABUCHI, M., "Microstructural and creep strength of welds in advanced ferritic power plant steels," *Science and Technology of Welding and Joining*, vol. 9, pp. 22 – 30, 2004.

[7] ALTENBACH, H., "Classical and nonclassical creep models," in *Creep and Damage in Materials and Structures* (ALTENBACH, H. and SKRZYPEK, J., eds.), pp. 45 – 95, Wien, New York: Springer, 1999. CISM Lecture Notes No. 399.

[8] ALTENBACH, H., "Creep analysis of thinwalled structures considering different structure mechanics models," in *Continuous Damage and Fracture* (BENALLAL, A., ed.), pp. 243 – 254, Paris et al.: Elsevier, 2000.

[9] ALTENBACH, H., "Consideration of stress state influences in the material modelling of creep and damage," in *IUTAM Symposium on Creep in Structures* (MURAKAMI, S. and OHNO, N., eds.), pp. 141 – 150, Dordrecht: Kluwer, 2001.

[10] ALTENBACH, H., "Creep analysis of thin-walled structures," *Z. Angew. Math. Mech.*, vol. 82, no. 8, pp. 507 – 533, 2002.

[11] ALTENBACH, H., ALTENBACH, J., and NAUMENKO, K., *Ebene Flächentragwerke*. Berlin: Springer, 1998.

[12] ALTENBACH, H., GORASH, Y., and NAUMENKO, K., "Steady-state creep of a pressurized thick cylinder in both the linear and the power law ranges," *Acta Mech.*, vol. 195, no. 1-4, pp. 263 – 274, 2008.

[13] ALTENBACH, H., HUANG, C., and NAUMENKO, K., "Creep damage predictions in thin-walled structures by use of isotropic and anisotropic damage models," *J. Strain Anal.*, vol. 37, no. 3, pp. 265 – 275, 2002.

[14] ALTENBACH, H., KUSHNEVSKY, V., and NAUMENKO, K., "On the use of solid- and shell-type finite elements in creep-damage predictions of thinwalled structures," *Arch. Appl. Mech.*, vol. 71, pp. 164 – 181, 2001.

[15] ALTENBACH, H., MORACHKOVSKY, O., NAUMENKO, K., and SYCHOV, A., "Geometrically nonlinear bending of thin-walled shells and plates under creep-damage conditions," *Arch. Appl. Mech.*, vol. 67, pp. 339 – 352, 1997.

[16] ALTENBACH, H., NAUMENKO, K., and GORASH, Y., "Numerical benchmarks for creep-damage modeling," *PAMM*, vol. 7, pp. 4040021 – 4040022, 2007.

[17] ALTENBACH, H., NAUMENKO, K., and GORASH, Y., "Creep analysis for a wide stress range based on stress relaxation experiments," *Modern Physics Letter B*, vol. 22, no. 31-32, pp. 5413 – 5418, 2008.

[18] ALTENBACH, J., ALTENBACH, H., and NAUMENKO, K., "Lebensdauerabschätzung dünnwandiger Flächentragwerke auf der Grundlage phänome-

nologischer Materialmodelle für Kriechen und Schädigung," *Technische Mechanik*, vol. 17, no. 4, pp. 353 – 364, 1997.

[19] ALTENBACH, J., ALTENBACH, H., and NAUMENKO, K., "Egde effects in moderately thick plates under creep damage conditions," *Technische Mechanik*, vol. 24, no. 3-4, pp. 254 – 263, 2004.

[20] ANSYS, Inc., Swanson Analysis Systems IP, Inc., *ANSYS Elements Reference*, Release 10.0 ed., 2005.

[21] ANSYS, Inc., Swanson Analysis Systems IP, Inc., *ANSYS, Inc. Theory Reference*, Release 10.0 ed., 2005.

[22] ANSYS, Inc., Swanson Analysis Systems IP, Inc., *Guide to ANSYS User Programmable Features*, Release 10.0 ed., 2005.

[23] ASBURY, F. E. and WILLOUGHBY, G., "Aging and creep behaviour of a Cr–Ni–Mn austenitic steel," in *Creep Strength in Steel and High-Temperature Alloys: Proc. of a Meeting held at the University of Sheffield on 20-22 Sep 1972* [97], pp. 144 – 151.

[24] ASHBY, M., SHERCLIFF, H., and CEBON, D., *Materials: Engineering, Science, Processing and Design*. New York et al.: Elsevier Ltd., 2007.

[25] ASHBY, M. F. and JONES, D. R. H., *Engineering Materials 1: An Introduction to Their Properties and Applications*. Oxford: Butterworth Heinemann, 2nd ed., 1996.

[26] AVEDJAN, A. and SCHJOKIN, I., "Proektirovanie svarnyh konstrukcij v Solidworks (Design of welded structures in Solidworks, in Russ.)," *SAPR and Graphics*, vol. 2, pp. 51 – 63, 2004.

[27] BAILEY, R. W., "Creep of steel under simple and compound stress," *Engineering*, vol. 121, pp. 129 – 265, 1930.

[28] BECKER, A. A. and HYDE, T. H., "Fundamental tests of creep behaviour," Report R0027, NAFEMS, Glasgow, Scotland, U.K., June 1993.

[29] BECKER, A. A., HYDE, T. H., SUN, W., and ANDERSSON, P., "Benchmarks for finite element analysis of creep continuum damage mechanics," *Computational Materials Science*, vol. 25, pp. 34 – 41, 2002.

[30] BECKER, A. A., HYDE, T. H., and XIA, L., "Numerical analysis of creep in components," *J. Strain Anal.*, vol. 29, no. 3, pp. 185 – 192, 1994.

[31] BECKITT, F. R., BANKS, T. M., and GLADMAN, T., "Secondary creep deformation in 18% Cr–10% Ni steel," in *Creep Strength in Steel and High-Temperature Alloys: Proc. of a Meeting held at the University of Sheffield on 20-22 Sep 1972* [97], pp. 71 – 77.

[32] BENDICK, W., HAARMANN, K., WELLNITZ, G., and ZSCHAU, M., "Eigenschaften der 9- bis 12%- Chromstähle und ihr Verhalten unter Zeitstandbeanspruchung," *VGB Kraftwerkstechnik*, vol. 73, no. 1, pp. 77 – 84, 1993.

[33] BETTEN, J., "Materialgleichungen zur Beschreibung des sekundären und tertiären Kriechverhaltens anisotroper Stoffe," *ZAMM*, vol. 64, pp. 211 – 220, 1984.

[34] BETTEN, J., *Creep Mechanics*. Berlin et al.: Springer-Verlag, 2005.

[35] BETTEN, J., BORRMANN, M., and BUTTERS, T., "Materialgleichungen zur Beschreibung des primären Kriechverhaltens innendruckbeanspruchter Zylinderschalen aus isotropem Werkstoff," *Ingenieur-Archiv*, vol. 60, no. 3, pp. 99 – 109, 1989.

[36] BETTEN, J. and BUTTERS, T., "Rotationssymmetrisches Kriechbeulen dünnwandiger Kreiszylinderschalen im primären Kriechbereich," *Forschung im Ingenieurwesen*, vol. 56, no. 3, pp. 84 – 89, 1990.

[37] BETTEN, J., EL-MAGD, E., MEYDANLI, S. C., and PALMEN, P., "Untersuchung des anisotropen Kriechverhaltens vorgeschädigter Werkstoffe am austenitischen Stahl X8CrNiMoNb 1616," *Arch. Appl. Mech.*, vol. 65, pp. 121 – 132, 1995.

[38] BIELECKI, M., KARCZ, M., RADULSKI, W., and BADUR, J., "Thermo-mechanical coupling betwee the flow of steam and deformation of the valve

during start-up of the 200 MW turbine," *Task Quarterly*, vol. 5, no. 2, pp. 125 – 140, 2001.

[39] BOLTON, J., "Analysis of structures based on a characteristic-strain model of creep," *Int. J. of Pressure Vessels & Piping*, vol. 85, pp. 108 – 116, 2008.

[40] BOYLE, J. T. and SPENCE, J., *Stress Analysis for Creep*. London: Butterworth, 1983.

[41] CANE, B. J., "The process controlling creep and creep fracture of 2Cr–1Mo steel," CEGB Report RD/LR 1979, Central Electricity Generating Board Research Laboratories, Leatherhead, UK, 1979.

[42] CANE, B. J. and BROWNE, R. J., "Representative stresses for creep deformation and failure of pressurized tubes and pipes," *Int. J. of Pressure Vessels & Piping*, vol. 10, no. 2, pp. 119 – 128, 1982.

[43] CELLA, A. and FOSSATI, C., "Experience feedback on damage evaluation and residual life prediction on chemical and petrochemical plants," in *High Temperature Structural Design* (LARSSON, L. H., ed.), no. 12 in ESIS Publication, pp. 437 – 454, London, UK: Mechanical Engineering Publications, 1992.

[44] CERJAK, H. and LETOFSKY, E., "The effect of welding on the properties of advanced 9-12%Cr steels," *Science and Technology*, vol. 1, no. 1, pp. 36 – 42, 1996.

[45] CHABOCHE, J., "Continuum damage mechanics a tool to describe phenomena before crack initiation," *Nuclear Engineering and Design*, vol. 64, pp. 233 – 247, 1981.

[46] CHABOCHE, J. L., "Constitutive equations for cyclic plasticity and cyclic viscoplasticity," *Int. J. Plasticity*, vol. 5, pp. 247 – 302, 1989.

[47] CHAN, W., MCQUEEN, R. L., PRINCE, J., and SIDEY, D., "Metallurgical experience with high temperature piping in Ontario Hydro," in *Service Experience in Operation Plants* (LARSSON, L. H., ed.), no. 21 in ASME PVP Division, pp. 97 – 105, New York, USA: ASME, 1991.

BIBLIOGRAPHY

[48] CHOUDHARY, B. K., PANIRAJ, C., RAO, K., and MANNAN, S. L., "Creep deformation behavior and kinetic aspects of 9Cr-1Mo ferritic steel," *ISIJ International*, vol. 41, pp. 73 – 80, 2001.

[49] CHOW, C. L. and WANG, J., "An anisotropic theory of elasticity for Continuum Damage Mechanics," *J. Fracture*, vol. 33, pp. 3 – 16, 1987.

[50] COBLE, R. I., "A model for boundary diffusion controlled creep in polycrystalline materials," *J. Appl. Phys.*, vol. 34, pp. 1679 – 1682, 1963.

[51] COCKS, A. C. F. and LECKIE, F. A., "Creep rupture of shell structures subjected to cyclic loading," *Trans. of ASME. J. Appl. Mech.*, vol. 55, pp. 294 – 298, 1988.

[52] COLLINS, M. J., "Creep strength in steel and high-temperature alloys: Proc. of a meeting held at the university of sheffield on 20-22 sep 1972," in *Creep Strength in Steel and High-Temperature Alloys: Proc. of a Meeting held at the University of Sheffield on 20-22 Sep 1972* [97], p. 217.

[53] CONTROL COMPONENTS INC., "Brochures: BTG valves product digest." World Wide Web portable document format (PDF), http://www.ccivalve.com/pdf/605.pdf, 2002.

[54] CONTROL COMPONENTS INC., "Data sheets: turbine quick stop valve (VQS), turbine flow control valve (VPC)." World Wide Web portable document format (PDF), http://www.ccivalve.com/pdf/552.pdf, 2003.

[55] CONTROL COMPONENTS INC., "Data sheets: turbine bypass stop valve (VS)." World Wide Web portable document format (PDF), http://www.ccivalve.com/pdf/615.pdf, 2005.

[56] CORDEBOIS, J. and SIDOROFF, F., "Damage induced elastic anisotropy," in *Mechanical Behaviours of Anisotropic Solids* (BOEHLER, J. P., ed.), pp. 761 – 774, Boston: Martinus Nijhoff Publishers, 1983.

[57] DA C. ANDRADE, E. N., "On the viscous flow of metals, and allied phenomena," *Proc. R. Soc. Lond.*, vol. ALXXXIV, pp. 1 – 12, 1910.

[58] DIMMLER, G., WEINERT, P., and CERJAK, H., "Investigations and analysis on the stationady creep behaviour of 9-12% chromium ferritic martensitic steels," in *Materials for Advances Power Engineering 2002: Proc. of the 7th Liège Conference* (LECOMTE-BECKERS, J., CARTON, M., SCHUBERT, F., and ENNIS, P. J., eds.), no. 21 in Reihe Energietechnik / Energy Technology, (Jülich, Germany), pp. 1539 – 1550, Institut für Werkstoffe und Verfahren der Energietechnik, Forschungszentrum Jülich GmbH, Sept. 2002.

[59] DIMMLER, G., WEINERT, P., and CERJAK, H., "Extrapolation of short-term creep rupture data – The potential risk of over-estimation," *Int. J. Pres. Ves. & Piping*, vol. 85, no. 1, pp. 55 – 62, 2008.

[60] DYSON, B. F., "Mechanical testing of high-temperature materials: modelling data-scatter," in *High-temperature structural materials* (CAHN, R. W., EVANS, A. G., and MCLEAN, M., eds.), pp. 160 – 192, London et al.: Chapman & Hall, 1996.

[61] DYSON, B. F. and MCLEAN, M., "Microstructural evolution and its effects on the creep performance of high temperature alloys," in *Microstructural Stability of Creep Resistant Alloys for High Temperature Plant Applications* (STRANG, A., CAWLEY, J., and GREENWOOD, G. W., eds.), pp. 371 – 393, Cambridge: Cambridge University Press, 1998.

[62] DYSON, B. F. and MCLEAN, M., "Micromechanism-quantification for creep constitutive equations," in *IUTAM Symposium on Creep in Structures* (MURAKAMI, S. and OHNO, N., eds.), pp. 3–16, Dordrecht: Kluwer, 2001.

[63] EARTHMAN, J. C., GIBELING, J. C., HAYES, R. W., and ET. AL., "Creep and stress-relaxation testing," in *Mechanical Testing and Evaluation* (KUHN, H. and MEDLIN, D., eds.), no. 8 in ASM Handbook, chapter 5, pp. 783 – 938, Materials Park, OH, USA: ASM International, 2000.

[64] EGGELER, G., "The effect of long term creep on particle coarsening in tempered martensite ferritic steels," *Acta Metallurgica*, vol. 37, no. 12, pp. 3225 – 3234, 1989.

[65] EGGELER, G., RAMETKE, A., COLEMAN, M., CHEW, B., PETER, G., BURBLIES, A., HALD, J., JEFFEREY, C., RANTALA, J., DEWITTE, M., and MOHRMANN, R., "Analysis of creep in a welded p91 pressure vessel," *Int. J. Pres. Ves. & Piping*, vol. 60, pp. 237 – 257, 1994.

[66] ELLIS, F. V. and TORDONATO, S., "Calculation of stress relaxation properties for type 422 stainless steel," *J. Pres. Vess. Tech.*, vol. 122, no. 1, pp. 66 – 71, 2000.

[67] FESSLER, H. and HYDE, T. H., "The use of model materials to simulate creep behavior," *J. Strain Anal.*, vol. 29, no. 3, pp. 193 – 200, 1994.

[68] FOLDYNA, V., JAKOBOVÁ, A., PRNKA, T., and SOBOTKA, J., "Creep strength in steel and high-temperature alloys: Proc. of a meeting held at the university of sheffield on 20-22 sep 1972," in *Creep Strength in Steel and High-Temperature Alloys: Proc. of a Meeting held at the University of Sheffield on 20-22 Sep 1972* [97], p. 230.

[69] FRANÇOIS, D., PINEAU, A., and ZAOU, A., *Comportement mécanique des matériaux: Viscoplasticité, endommagement, mécanique de la rupture, mécanique du contact*. Paris: Hermès, 1993.

[70] FROST, H. J. and ASHBY, M. F., *Deformation Mechanism Maps*. Oxford: Pergamon Press, 1982.

[71] GAFFARD, V., BESSON, J., and GOURGUES-LORENZON, A. F., "Creep failure model of a tempered martensitic stainless steel integraiting multiple deformation and damage mechanisms," *Int. J. Fracture*, vol. 133, pp. 139 – 166, 2005.

[72] GAFFARD, V., GOURGUES-LORENZON, A. F., and BESSON, J., "High temperature creep flow and damage properties of the weakest area of 9Cr1Mo-NbW martensitic steel weldments," *ISIJ International*, vol. 45, no. 12, pp. 1915 – 1924, 2005.

[73] GAROFALO, F., *Fundamentals of Creep and Creep-Rupture in Metals*. New York: The Macmillan Co., 1965.

[74] GAUDIG, W., KUSSMAUL, K., MAILE, K., TRAMER, M., GRANACHER, J., and KLOOS, K. H., "A microstructural model to predict the multiaxial creep and damage in 12Cr Steel Grade at 550°C," in *Microstructure and Mechanical Properties of Metallic High-Temperature Materials: Research Report/DFG* (MUGHARBI, H., GOTTSTEIN, G., MECKING, H., RIEDEL, H., and TOBOLSKI, J., eds.), pp. 192 – 205, Weinheim: Wiley-VCH, 1999.

[75] GHOSH, R. and MCLEAN, M., "High temperature deformation in engineering alloys-modelling for strain of load control," *Acta Metall. Mater.*, vol. 40, pp. 3075 – 3083, 1992.

[76] GOOCH, D. J., "Remnant creep life prediction in ferritic materials," in *Comprehensive Structural Integrity, Vol. 5, Creep and High-Temperature Failure* (SAXENA, A., ed.), pp. 309–359, Amsterdam et al.: Elsevier, 2003.

[77] GORASH, E., LVOV, G., HARDER, J., KOSTENKO, Y., and WIEGHARDT, K., "Comparative analysis of the creep behaviour in a power plant component using different material models," in *Creep & Fracture in High Temperature Components: Design & Life Assessment Issues* (SHIBLI, I. A., HOLDSWORTH, S. R., and MERCKLING, G., eds.), Proc. of the ECCC Creep Conf., (London, UK), pp. 573 – 581, ECCC, DESTech Publications, Inc., Sept. 2005.

[78] GORASH, Y., ALTENBACH, H., and NAUMENKO, K., "Modeling of primary and secondary creep for a wide stress range," *PAMM*, vol. 8, pp. 10207 – 10208, 2008.

[79] GORASH, Y. M., "Zastosuvannja izotropnoi ta anizotropnoi koncepcij poshkodzhuvannosti do rozrahunku tryvaloi micnosti rotoru parovoi turbiny v umovah vysokotemperaturnoi povzuchosti (Application of isotropic and anisotropic concepts of damage to the long-term strength analysis of the steam turbine rotor in high-temperature creep conditions, in Ukr.)," in *Trans. of NTU "KhPI"* (MORACHKOVSKIY, O. K., ed.), no. 21 in Dynamics and Strength of Machines, pp. 92 – 101, Kharkiv, Ukraine: NTU "KhPI" Publishing Center, 2006.

[80] GORASH, Y. N., LVOV, G. I., NAUMENKO, K. V., and ALTENBACH, H., "Analiz vysokotemperaturnoj polzuchesti T-obraznogo soedinenija dvuh trub, nagruzhennyh davleniem, s uchjotom anizotropii svojstv materiala v zone svarnogo shva (High-temperature creep analysis of pressurized T-piece pipe weldment considering anisotropic material properties in a welding zone, in Russ.)," in *Trans. of NTU "KhPI"* (MORACHKOVSKIY, O. K., ed.), no. 20 in Dynamics and Strength of Machines, pp. 57 – 66, Kharkiv, Ukraine: NTU "KhPI" Publishing Center, 2005.

[81] GUMMERT, P., "General constitutive equations for simple and non–simple materials," in *Creep and Damage in Materials and Structures* (ALTENBACH, H. and SKRZYPEK, J., eds.), pp. 1 – 43, Wien, New York: Springer, 1999. CISM Lecture Notes No. 399.

[82] HARPER, J. G. and DORN, J. E., "Viscous creep of aluminum near its melting temperature," *Acta Metallurgica*, vol. 5, no. 11, pp. 654 – 665, 1957.

[83] HARPER, J. G., SHEPARD, L. A., and DORN, J. E., "Creep of aluminum under extremely small stresses," *Acta Metallurgica*, vol. 6, pp. 509 – 518, 1958.

[84] HAUPT, P., *Continuum Mechanics and Theory of Materials*. Berlin: Springer, 2002.

[85] HAYHURST, D. R., "Creep rupture under multiaxial states of stress," *J. Mech. Phys. Solids*, vol. 20, pp. 381 – 390, 1972.

[86] HAYHURST, D. R., "The use of Continuum Damage Mechanics in creep analysis for design," *J. Strain Anal.*, vol. 25, no. 3, pp. 233 – 241, 1994.

[87] HAYHURST, D. R., "Materials data bases and mechanisms-based constitutive equations for use in design," in *Creep and Damage in Materials and Structures* (ALTENBACH, H. and SKRZYPEK, J., eds.), pp. 285 – 348, Wien, New York: Springer, 1999. CISM Lecture Notes No. 399.

[88] HAYHURST, D. R., "Computational Continuum Damage Mechanics: its use in the prediction of creep fracture in structures - past, present and future," in

IUTAM Symposium on Creep in Structures (MURAKAMI, S. and OHNO, N., eds.), pp. 175 – 188, Dordrecht: Kluwer, 2001.

[89] HAYHURST, D. R. and LECKIE, F. A., "High temperature Creep Continuum Damage in metals," in *Yielding, Damage and Failure of Anisotropic Solids* (BOEHLER, J. P., ed.), pp. 445 – 464, London: Mechanical Engineering Publications, 1990.

[90] HAYHURST, D. R., WONG, M. T., and VAKILI-TAHAMI, F., "The use of CDM analysis techniques in high temperature creep failure of welded structures," *JSME Int. J. Series A*, vol. 45, pp. 90 – 97, 2002.

[91] HERRING, C., "Diffusional viscosity of a polycrystalline solid," *J. Appl. Phys.*, vol. 21, no. 5, pp. 437 – 445, 1950.

[92] HOLDSWORTH, S. R. and MERCKLING, G., "ECCC developments in the assessment of creep-rupture properties," in *Parsons 2003 – Engineering Issues in Turbine Machinery, Power Plant and Renewables* (STRANG, A., CONROY, R. D., BANKS, W. M., BLACKLER, M., LEGGETT, J., MCCOLVIN, G. M., SIMPLSON, S., SMITH, M., STARR, F., and VANSTONE, R. W., eds.), no. Sept. 16-18, 2003 in Proc. of the 6th Inter. Charles Parsons Turbine Conf., (London, U.K.), pp. 411 – 426, Maney Publishing, 2003.

[93] HULT, J. A., *Creep in Engineering Structures*. Waltham: Blaisdell Publishing Company, 1966.

[94] HYDE, T. H., SUN, W., AGYAKWA, P. A., SHIPWAY, P. H., and WILLIAMS, J. A., "Anisotropic creep and fracture behaviour of a 9CrMoNbV weld metal at 650°C," in *Anisotropic Behaviour of Damaged Materials* (SKRZYPEK, J. J. and GANCZARSKI, A. W., eds.), pp. 295–316, Berlin et al.: Springer, 2003.

[95] HYDE, T. H., SUN, W., and WILLIAMS, J. A., "Creep analysis of pressurized circumferential pipe weldments - a review," *J. Strain Anal.*, vol. 38, no. 1, pp. 1 – 29, 2003.

[96] HYDE, T. H., XIA, L., and BECKER, A. A., "Prediction of creep failure in aeroengine materials under multi-axial stress states," *Int. J. Mech. Sci.*, vol. 38, no. 4, pp. 385 – 403, 1996.

BIBLIOGRAPHY

[97] Iron and Steel Institute, *Creep Strength in Steel and High-Temperature Alloys: Proc. of a Meeting held at the University of Sheffield on 20-22 Sep 1972*, (London, UK), Metals Society, 1974.

[98] JIANPING, J., GUANG, M., YI, S., and SONGBO, X., "An effective continuum damage mechanics model for creep–fatigue life assessment of a steam turbine rotor," *Int. J. of Pressure Vessels & Piping*, vol. 80, pp. 389 – 396, 2003.

[99] JOHNSON, A. E., HENDERSON, J., and KHAN, B., *Complex Stress and Creep Relaxation and Fracture of Metallic Alloys*. Edinburgh: HMSO, 1962.

[100] KACHANOV, L. M., "O vremeni razrusheniya v usloviyakh polzuchesti (On the time to rupture under creep conditions, in Russ.)," *Izv. AN SSSR. Otd. Tekh. Nauk*, no. 8, pp. 26 – 31, 1958.

[101] KERN, T.-U., WIEGHARDT, K., and KIRCHNER, H., "Materials and design solutions for advanced steam power plants," in *Advances in Materials Technology for Fossil Power Plants: Proc. of the 4th International Conference* (VISWANATHAN, R., GANDY, D., and COLEMAN, K., eds.), (Materials Park, OH, USA), pp. 20 – 34, ASM International, Sept. 2004.

[102] KIMURA, K., "9Cr-1Mo-V-Nb Steel," in *Creep Properties of Heat Resistant Steels and Superalloys* (YAGI, K., MERCKLING, G., KERN, T.-U., IRIE, H., and WARLIMONT, H., eds.), no. VIII/2B in Numerical Data and Functional Relationships in Science and Technology, pp. 126 – 133, Berlin et al.: Springer-Verlag, 2006.

[103] KIMURA, K., KUSHIMA, H., ABE, F., and YAGI, K., "Evaluation of the creep strength property from a viewpoint of inherent creep strength for ferritic creep resistant steels," in *Microstructural Stability of Creep Resistant Alloys for High Temperature Plant Applications* (STRANG, A., CAWLEY, J., and GREENWOOD, G. W., eds.), no. 2 in Microstructure of High Temperature Materials, pp. 185 – 196, Cambridge: Cambridge University Press, 1998.

[104] KLENK, A., SCHEMMEL, J., and MAILE, K., "Numerical modelling of ferritic welds and repair welds," *OMMI*, vol. 2, no. 2, pp. 1 – 14, 2003. Available from http://www.ommi.co.uk.

[105] KLOC, L. and FIALA, J., "On creep behaviour of several metallic materials at low stresses and elevated temperatures," *Chemical Papers*, vol. 53, no. 3, pp. 155 – 164, 1999.

[106] KLOC, L. and FIALA, J., "Viscous creep in metals at intermediate temperatures," *Kovové Materiály*, vol. 43, no. 2, pp. 105 – 112, 2005.

[107] KLOC, L. and SKLENIČKA, V., "Transition from power-law to viscous creep behaviour of P-91 type heat-resistant steel," *Mater. Sci. & Eng.*, vol. A234-A236, pp. 962 – 965, 1997.

[108] KLOC, L. and SKLENIČKA, V., "Confirmation of low stress creep regime in 9% Chromium steel by stress change creep experiments," *Mater. Sci. & Eng.*, vol. A387-A389, pp. 633 – 638, 2004.

[109] KLOC, L., SKLENIČKA, V., DLOUHÝ, A. A., and KUCHAŘOVÁ, K., "Power-law and viscous creep in advanced 9%Cr steel," in *Microstructural Stability of Creep Resistant Alloys for High Temperature Plant Applications* (STRANG, A., CAWLEY, J., and GREENWOOD, G. W., eds.), no. 2 in Microstructure of High Temperature Materials, pp. 445 – 455, Cambridge: Cambridge University Press, 1998.

[110] KLOC, L., SKLENIČKA, V., and VENTRUBA, J., "Comparison of low creep properties of ferritic and austenitic creep resistant steels," *Mater. Sci. & Eng.*, vol. A319-A321, pp. 774 – 778, 2001.

[111] KLUEH, R. L., "Elevated-temperature ferritic and martensitic steels and their application to future nuclear reactors," ORNL Report TM-2004/176, Oak Ridge National Laboratory, Metals and Ceramics Division, Oak Ridge, Tennessee, Nov. 2004.

[112] KONKIN, V. N. and MORACHKOVSKII, O. K., "Creep and long-term strength of light alloys with anisotropic properties," *Strength of Materials*, vol. 19, no. 5, pp. 626 – 631, 1987.

[113] KOSTENKO, Y., "Development of advanced method for creep and damage estimation of turbine components," Report SP No.410294, Siemens AG R&D, Mülheim/Ruhr, Germany, 2005.

[114] KOSTENKO, Y., LVOV, G., GORASH, E., ALTENBACH, H., and NAUMENKO, K., "Power plant component design using creep-damage analysis," in *Proc. of Int. Mech. Eng. Conf. & Exp. "IMECE-2006"*, no. 13710 in IMECE, (Chicago, USA), pp. 1 – 10, ASME International, ASME, Nov. 2006.

[115] KOUNDY, V., FORGERON, T., and NAOUR, F. L., "Modeling of multiaxial creep behavior for incoloy 800 tubes under internal pressure," *Trans. ASME J. of Pressure Vessel Technology*, vol. 119, pp. 313 – 318, 1997.

[116] KOWALEWSKI, Z. L., HAYHURST, D. R., and DYSON, B. F., "Mechanisms-based creep constitutive equations for an aluminium alloy," *J. Strain Anal.*, vol. 29, no. 4, pp. 309 – 316, 1994.

[117] KRIEG, R., "Reactor pressure vessel under severe accident loading," Final report of EU-project: Contract FI4S-CT95-0002, Forschungszentrum Karlsruhe GmbH, Karlsruhe, Germany, 1999.

[118] KUSHIMA, K., KIMURA, K., and ABE, F., "Long-term creep strength prediction of high Cr ferritic creep resistant steels," in *Materials for Advances Power Engineering 2002: Proc. of the 7th Liège Conference* (LECOMTE-BECKERS, J., CARTON, M., SCHUBERT, F., and ENNIS, P. J., eds.), no. 21 in Reihe Energietechnik / Energy Technology, (Jülich, Germany), pp. 1581 – 1590, Institut für Werkstoffe und Verfahren der Energietechnik, Forschungszentrum Jülich GmbH, Sept. 2002.

[119] LANGDON, T. G., "Creep at low stresses: an evaluation of diffusion creep and Harper–Dorn creep as viable creep mechanisms," *Metall. & Mater. Trans.*, vol. 33, no. A, pp. 249 – 259, 2002.

[120] LECKIE, F. and HAYHURST, D., "Creep rupture of structures," *Proc. R. Soc.*, vol. 340, pp. 323 – 347, 1974.

[121] LECKIE, F. A. and HAYHURST, D. R., "Constitutive equations for creep rupture," *Acta Metall.*, vol. 25, pp. 1059 – 1070, 1977.

[122] LEE, J. S., ARMAKI, H. G., MARUYAMA, K., MURAKI, T., and ASAHI, H., "Causes of breakdown of creep strength in 9Cr-1.8W-0.5Mo-VNb steel," *Mater. Sci. & Eng.*, vol. A428, pp. 270 – 275, 2006.

[123] LEMAITRE, J., "Evaluation of dissipation and damage in metals submitted to dynamic loading," in *Proc. of the 1st International Conference on Mechanical Behavior of Materials*, no. 1 in ICM, (Kyoto, Japan), 1971.

[124] LEMAITRE, J., *A Course on Damage Mechanics*. Berlin: Springer, 1996.

[125] LEMAITRE, J., ed., *Handbook of Materials Behavior Models*. San Diego: Academic Press, 2001.

[126] LEMAITRE, J. and CHABOCHE, J.-L., *Mechanics of Solid Materials*. Cambridge: Cambridge University Press, 1990.

[127] LEVITIN, V., *High Temperature Strain of Metals and Alloys*. Weinheim: Wiley-VCH, 2006.

[128] LIFSHITZ, I. M., "On the theory of diffusion-viscous flow of polycrystalline bodies," *Soviet Physics. Journal of Experimental and Theoretical Physics*, vol. 17, pp. 909 – 920, 1963.

[129] LVOV, G. I., LYSENKO, S. V., and GORASH, Y. N., "Neizotermicheskaja polzuchest i povrezhdaemost elementov parovyh turbin (Non-isothermal creep and damage of steam turbine components, in Russ.)," in *Trans. of NTU "KhPI"* (MORACHKOVSKIY, O. K., ed.), no. 21 in Dynamics and Strength of Machines, pp. 75 – 88, Kharkiv, Ukraine: NTU "KhPI" Publishing Center, 2006.

[130] LVOV, G. I., LYSENKO, S. V., and GORASH, Y. N., "Dlitelnaja prochnost klapana vysokogo davlenija s uchjotom neodnorodnogo raspredelenija temperatury (Long-term strength of the high-pressure valve considering a non-uniform temperature distribution, in Russ.)," in *Trans. of NTU "KhPI"* (MORACHKOVSKIY, O. K., ed.), no. 22 in Dynamics and Strength of Machines, pp. 98 – 106, Kharkiv, Ukraine: NTU "KhPI" Publishing Center, 2007.

BIBLIOGRAPHY

[131] LVOV, G. I., LYSENKO, S. V., and GORASH, Y. N., "Polzuchest i dlitelnaja prochnost komponentov gazovyh turbin s uchjotom neodnorodnogo raspredelenija temperatur (The creep and long-term strength of components of gas turbines considering a non-uniform distribution of temperatures, in Russ.)," in *Problems of Dynamics and Strength in Gas-Turbine Manufacture (GTD-2007)*, Proc. 3rd Int. Sci. & Tech. Conf., (Kyiv, Ukraine), National Academy of Sciences Ukraine, Institute of Strength Problems, May 2007. 8 p., submitted.

[132] LVOV, G. I., LYSENKO, S. V., and GORASH, Y. N., "Prikladnye zadachi neizotermicheskoj teorii dlitelnoj prochnosti (Applied problems of unisothermal theory of the long-term strength, in Russ.)," in *Problems of Computational Mechanics and Strength of Structures: Transactions*, no. 11 in Dnipropetrovsk National University, pp. 71 – 78, Dnipropetrovsk, Ukraine: Science and Education, Oct. 2007.

[133] MAILE, K., KLENK, A., ROOS, E., HUSEMANN, R. U., and HELMRICH, A., "Development and qualification of new boiler and piping materials for high efficiency USC plants," in *Advances in Materials Technology for Fossil Power Plants: Proc. of the 4th International Conference* (VISWANATHAN, R., GANDY, D., and COLEMAN, K., eds.), (Materials Park, OH, USA), pp. 152 – 164, ASM International, Sept. 2004.

[134] MALININ, N. N., *Prikladnaya teoriya plastichnosti i polzuchesti (Applied theory of plasticity and creep, in Russ.)*. Moskva: Mashinostroenie, 1975.

[135] MALININ, N. N., *Raschet na polzuchest' konstrukcionnykh elementov (Creep calculations of structural elements, in Russ.)*. Moskva: Mashinostroenie, 1981.

[136] MASOOD, Z., ARORA, P. R., and BASRI, S. N., "A modified approach for life assessment of high temperature power plant components," in *Proc. of the World Engineering Congress 99 & Exhibition (WEC'99)*, (Kuala Lumpur, Malaysia), pp. 23 – 28, University Putra Malaysia, July 1999.

[137] MCCOY, H. E., "Creep properties of selected refractory alloys for potential space nuclear power applications," ORNL Report TM-10127, Oak Ridge Na-

tional Laboratory, Metals and Ceramics Division, Oak Ridge, Tennessee, Oct. 1986.

[138] MCLAUCHLIN, I. R., "Temperature dependence of steady-state creep in 20%Cr–25%Ni–Nb stabilized stainless steel," in *Creep Strength in Steel and High-Temperature Alloys: Proc. of a Meeting held at the University of Sheffield on 20-22 Sep 1972* [97], pp. 86 – 90.

[139] MERCKLING, G., "Long-term creep rupture strength assessment: Development of the european collaborative creep committee post-assessment tests," *Int. J. of Pressure Vessels & Piping*, vol. 85, no. 1, pp. 2 – 13, 2008.

[140] MOBERG, F., "Implementation of constitutive equations for creep damage mechanics into the ABAQUS finite element code – Subroutine UMAT," Report 95/05, SAQ/R&D, Stockholm, Sweden, 1995.

[141] MURAKAMI, S., "A continuum mechanics theory of anisotropic damage," in *Yielding, damage and failure of anisotropic solids* (BOEHLER, J. P., ed.), pp. 465 – 482, London, U.K.: Mechanical Engineering Publications, 1990.

[142] MURAKAMI, S. and OHNO, N., "A continuum theory of creep and creep damage," in *Creep in Structures* (PONTER, A. R. S. and HAYHURST, D. R., eds.), pp. 422 – 444, Berlin: Springer, 1981.

[143] MURAKAMI, S. and SANOMURA, Y., "Creep and creep damage of copper under multiaxial states of stress," in *Plasticity Today – Modelling, Methods and Applications* (SAWCZUK, A. and BIANCHI, B., eds.), pp. 535 – 551, London, New York: Elsevier, 1985.

[144] NABARRO, F. R. N., "Deformation of crystals by the motion of single ions," in *Report of a Conference on the Strength of Solids (Bristol, 7th-9th July 1947)* (MOTT, N. F., ed.), no. 38 in Strength of solids, (London, U.K.), pp. 75 – 90, The Physical Society, 1948.

[145] NABARRO, F. R. N. and DE VILLIERS, H. L., *The Physics Of Creep. Creep and Creep-resistant Alloys*. London: Taylor & Francis Ltd., 1995.

[146] NAUMENKO, K., *Modellierung und Berechnung der Langzeitfestigkeit dünnwandiger Flächentragwerke unter Berücksichtigung von Werkstoffkriechen und Schädigung*. Dissertation, Fakultät für Maschinenbau, Otto-von-Guericke-Universität, Magdeburg, Germany, 1996.

[147] NAUMENKO, K., *Modeling of High-Temperature Creep for Structural Analysis Applications*. Habilitationschrift, Mathematisch-Naturwissenschaftlich-Technische Fakultät, Martin-Luther-Universität Halle-Wittenberg, 2006.

[148] NAUMENKO, K., "Modellierung des Hochtemperaturkriechens für strukturmechanische Anwendungen." Vortrag im Rahmen des Habilitationsverfahrens, Feb. 20, 2006.

[149] NAUMENKO, K. and ALTENBACH, H., "Ein phänomenologisches Modell für anisotropes Kriechen in Schweißnähten," *PAMM*, vol. 4, pp. 227 – 228, 2004.

[150] NAUMENKO, K. and ALTENBACH, H., "Ein phänomenologisches Modell für anisotropes Kriechen in Schweißnähten." Vortrag im Rahmen der GAMM Jahrestagung, Mar. 24, 2004.

[151] NAUMENKO, K. and ALTENBACH, H., "A phenomenological model for anisotropic creep in a multi-pass weld metal," *Arch. Appl. Mech.*, vol. 74, pp. 808 – 819, 2005.

[152] NAUMENKO, K. and ALTENBACH, H., *Modelling of Creep for Structural Analysis*. Berlin et al.: Springer-Verlag, 2007.

[153] NAUMENKO, K., ALTENBACH, H., and GORASH, Y., "Creep analysis with a stress range dependent constitutive model," *Arch. Appl. Mech.*, vol. 79, no. 6-7, pp. 619 – 630, 2009.

[154] NEUBAUER, B. and WEDEL, U., "Restlife estimation of creeping components by means of replicas," in *Advances in Life Prediction Methods: Proc. of the ASME International Conference* (WOODFORD, D. A. and WHITEHEAD, J. R., eds.), (New York), pp. 307 – 314, ASME, 1983.

[155] NIU, L., KOBAYASHI, M., and TAKAKU, H., "Creep rupture properties of an austenitic steel with high ductility under multi-axial stresses," *ISIJ International*, vol. 42, no. 10, pp. 1156 – 1161, 2002.

[156] NIX, W. D., EARTHMAN, J. C., EGGELER, G., and ILSCHNER, B., "The principal facet stress as a parameter for predicting creep rupture under multi-axial stresses," *Acta Metallurgica*, vol. 37, no. 4, pp. 1067 – 1077, 1989.

[157] NORTON, F. H., *The Creep of Steel at High Temperature*. New York: McGraw-Hill Book Co., 1929.

[158] NOWAK, G. and RUSIN, A., "Lifetime deterioration of turbine components during start-ups," *OMMI*, vol. 3, no. 1, pp. 1 – 10, 2004. Available from http://www.ommi.co.uk.

[159] ODQVIST, F. K. G., *Mathematical Theory of Creep and Creep Rupture*. Oxford: Oxford University Press, 1974.

[160] ODQVIST, F. K. G., "Historical survey of the development of creep mechanics from its beginnings in the last century to 1970," in *Creep in Structures* (PONTER, A. R. S. and HAYHURST, D. R., eds.), pp. 1 – 12, Berlin et al.: Springer, 1981.

[161] ODQVIST, F. K. G. and HULT, J., *Kriechfestigkeit metallischer Werkstoffe*. Berlin u.a.: Springer, 1962.

[162] OKADA, M., "Data sheets on the elevated-temperature properties of 9Cr-1Mo-V-Nb steel tubes for boilers and heat exchangers (ASME SA-213/SA-213M Grade T91) and 9Cr-1Mo-V-Nb steel plates for boilers and pressure vessels (ASME SA-387/SA-387M Grade 91)," NRIM Creep Data Sheet 43, National Research Institute for Metals, Tokyo, Japan, Sept. 1996.

[163] OKADA, M., "Data sheets on the elevated-temperature stress relaxation properties of 1Cr-0.5Mo-0.25V steel and 12Cr-1Mo-1W-0.25V steel bolting materials for high temperature service," NRIM Creep Data Sheet 44, National Research Institute for Metals, Tokyo, Japan, Mar. 1997.

BIBLIOGRAPHY

[164] OKADA, M., "Data sheets on the elevated-temperature properties of 12Cr-1Mo-1W-0.3V heat-resisting steel bars for turbine blades (SUH 616-B)," NRIM Creep Data Sheet 10B, National Research Institute for Metals, Tokyo, Japan, Mar. 1998.

[165] ORLOVÁ, A., BURŠÍK, J., KUCHAŘOVÁ, K., and SKLENIČKA, V., "Microstructural development during high temperature creep of 9%Cr steel," *Mater. Sci. & Eng.*, vol. A245, pp. 39 – 48, 1998.

[166] PENNY, R. K. and MARIOTT, D. L., *Design for Creep*. London: Chapman & Hall, 1995.

[167] PERRIN, I. J. and HAYHURST, D. R., "Creep constitutive equations for a 0.5Cr-0.5Mo-0.25V ferritic steel in the temperature range 600-675°C," *J. Strain Anal.*, vol. 31, no. 4, pp. 299 – 314, 1996.

[168] PODGORNY, A. N., BORTOVOJ, V. V., GONTAROVSKY, P. P., KOLOMAK, V. D., LVOV, G. I., MATYUKHIN, Y. J., and MORACHKOVSKY, O. K., *Polzuchest' elementov mashinostroitel'nykh konstrykcij (Creep of mashinery structural members, in Russ.)*. Kiev: Naukova dumka, 1984.

[169] POLCIK, P., SAILER, T., BLUM, W., STRAUB, S., BURŠIK, J., and ORLOVÁ, A., "On the microstructural developmenr of the tempered martensitic Cr-steel P91 during long-term creep – a comparison of data," *Mater. Sci. & Eng.*, vol. A260, pp. 252 – 259, 1999.

[170] GERMAN INSTITUTE FOR STANDARDIZATION, *Welding – Basic weld joint details in steel – Part 2: Steel structures, in Germ.* No. 8558-10 in E DIN, Berlin, Germany: Deutsches Institut für Normung e.V., 1997.

[171] STATE STANDARD OF USSR, *Welded joints in steel pipelines. Main types, design elements and dimensions, in Russ.* No. 16037-80 in GOST, Moscow, Russian Federation: State Committee of USSR on Standards, 1980.

[172] QI, W. and BROCKS, W., "ABAQUS user subroutines for the simulation of viscoplastic behaviour including anisotropic damage for isotropic materials and for single crystals," Report WMS/01/5, Institut für Werkstofforschung, GKSS-Forschungszentrum, Geesthacht, Germany, June 2001.

[173] RABOTNOV, Y. N., "O mechanizme dlitel'nogo razrusheniya (A mechanism of the long term fracture, in Russ.)," *Voprosy prochnosti materialov i konstruktsii, AN SSSR*, pp. 5 – 7, 1959.

[174] RABOTNOV, Y. N., *Creep Problems in Structural Members*. Amsterdam: North-Holland, 1969.

[175] RAJ, S. V., "Creep and fracture of dispersion-strengthened materials," NASA Contractor Report 185299, National Aeronautics and Space Administration, Lewis Research Center, Cleveland, Ohio, June 1991.

[176] RAJ, S. V., ISKOVITZ, I. S., and FREED, A. D., "Modeling the role of dislocation substructure during class M and exponential creep," NASA Technical Memorandum 106986, National Aeronautics and Space Administration, Lewis Research Center, Cleveland, Ohio, Aug. 1995.

[177] RIEDEL, H., *Fracture at High Temperatures*. New York et al.: Springer-Verlag, 1987.

[178] RIETH, M., "A comprising steady-state creep model for the austenitic AISI 316 L(N) steel," *Journal of Nuclear Materials*, vol. 360-370, no. 2, pp. 915 – 919, 2007.

[179] RIETH, M., FALKENSTEIN, A., GRAF, P., HEGER, S., JÄNTSCH, U., KLIMIANKOU, M., MATERNA-MORRIS, E., and ZIMMERMANN, H., "Creep of the austenitic steel AISI 316 L(N): experiments and models," Report FZKA 7065, Institut für Materialforschung, Forschungszentrum Karlsruhe GmbH, Karlsruhe, Germany, Nov. 2004.

[180] ROBERTSON, D. G. and HOLDSWORTH, S. R., "Rupture strength, creep strength and relaxation strength values for carbon-manganese, low alloy ferritic, high alloy ferritic and austenitic steels, and high temperature bolting steels/alloys," ECCC Data Sheets 2, European Creep Collaborative Committee, European Technology Development Ltd, Ashtead, Surrey, UK, Sept. 2005.

[181] RÖSLER, J., HARDERS, H., and BÄKER, M., *Mechanical Behaviour of Engineering Materials*. Berlin et al.: Springer-Verlag, 2007.

[182] SDOBYREV, V. P., "Kriterij dlitel'noj prochnosti dlya nekotorykh zharoprochnykh splavov pri slozhnom napryazhennom sostoyanii (Criterion of long term strength of some high-temperature alloys under multiaxial stress state, in Russ.)," *Izv. AN SSSR. OTN. Mekh. i Mashinostroenie*, vol. , no. 6, pp. 93 – 99, 1959.

[183] SEGLE, P., SAMUELSON, L. Å., ANDERSSON, P., and MOBERG, F., "Inelasticity and damage in solids subject to microstructural change," in *Creep & Fracture in High Temperature Components: Design & Life Assessment Issues* (JORDAAN, I. J., SESHADRI, R., and MEGLIS, I. L., eds.), no. Sept. 25-27, 1996 in Proc. of the The Lazar M. Kachanov Symp., (Newfoundland, Canada), Memorial University of Newfoundland, 1997.

[184] SEGLE, P., TU, S. T., STORESUND, J., and SAMUELSON, L. A., "Some issues in life assessment of longitudinal seam welds based on creep tests with cross-weld specimens," *Int. J. of Pressure Vessels & Piping*, vol. 66, pp. 199 – 222, 1996.

[185] SHERBY, O. D. and BURKE, P. M., "Mechanical behavior of crystalline solids at elevated temperature," *Prog. Mater. Sci.*, vol. 13, no. 7, pp. 325 – 390, 1967.

[186] SHIBLI, I. A., "Performance of P91 thick section welds under steady and cyclic loading conditions: power plant and research experience," *OMMI*, vol. 1, no. 3, pp. 1 – 17, 2002. Available from http://www.ommi.co.uk.

[187] SKLENIČKA, V., KUCHAŘOVÁ, K., KLOC, L., and SVOBODA, M., "Degradation processes in creep of 9-12%Cr ferritic steels," in *Advances in Materials Technology for Fossil Power Plants: Proc. of the 4th International Conference* (VISWANATHAN, R., GANDY, D., and COLEMAN, K., eds.), (Materials Park, OH, USA), pp. 1086 – 1100, ASM International, Sept. 2004.

[188] SKLENIČKA, V., KUCHAŘOVÁ, K., KLOC, L., SVOBODA, M., and STAUBLI, M., "The effect of microstructural stability on long-term creep behaviour of 9-12%Cr steels," in *Materials for Advances Power Engineering 2002: Proc. of the 7th Liège Conference* (LECOMTE-BECKERS, J., CARTON,

M., SCHUBERT, F., and ENNIS, P. J., eds.), no. 21 in Reihe Energietechnik / Energy Technology, (Jülich, Germany), pp. 1189 – 1200, Institut für Werkstoffe und Verfahren der Energietechnik, Forschungszentrum Jülich GmbH, Sept. 2002.

[189] SKLENIČKA, V., KUCHAŘOVÁ, K., KUDRMAN, J., SVOBODA, M., and KLOC, L., "Microstructure stability and creep behaviour of advanced high chromium ferritic steels," *Kovové Materiály*, vol. 43, no. 1, pp. 20 – 33, 2005.

[190] SKLENIČKA, V., KUCHAŘOVÁ, K., SVOBODA, M., KLOC, L., BURŠIK, J., and KROUPA, A., "Long term creep behavior of 9–12%Cr power plant steels," *Mater. Charact.*, vol. 51, no. 1, pp. 35 – 48, 2003.

[191] SKRZYPEK, J. and GANCZARSKI, A., *Modelling of Material Damage and Failure of Structures*. Foundation of Engineering Mechanics, Berlin: Springer, 1998.

[192] SKRZYPEK, J. J., *Plasticity and Creep*. Boca Raton: CRC Press, 1993.

[193] SŁUŻALEC, A., *Introduction to Nonlinear Thermomechanics*. Berlin: Springer, 1992.

[194] TANAKA, C. and OHBA, T., "Analysis of reloading sress relaxation behavior with specified reloading time intervals for high temperature bolting steels," *Transactions of National Research Institute for Metals*, vol. 26, no. 1, pp. 1 – 20, 1984.

[195] THREADGILL, P. L. and WILSHIRE, B., "Mechanisms of transient and steady-state creep in a γ'-hardened austenitic steel," in *Creep Strength in Steel and High-Temperature Alloys: Proc. of a Meeting held at the University of Sheffield on 20-22 Sep 1972* [97], pp. 8 – 14.

[196] VISWANATHAN, R., "Effect of stress and temperature on the creep and rupture behavior of a 1.25% chromium – 0.5 % molybdenum steel," *Metall. Trans. A*, vol. 8, no. 6, pp. 877 – 884, 1977.

[197] VISWANATHAN, R., *Damage Mechanisms and Life Assessment of High-Temperature Components*. Metals Park, Ohio: ASM International, 1989.

BIBLIOGRAPHY

[198] WALLES, K. F. A. and GRAHAM, A., "On the extrapolation and scatter of creep data," NGTE Report R247, National Gas Turbine Establishment, 1961.

[199] WEBER, J., KLENK, A., and RIEKE, M., "A new method of strength calculation and lifetime prediction of pipe bends operating in the creep range," *Int. J. of Pressure Vessels & Piping*, vol. 82, pp. 77 – 84, 2005.

[200] WOLBERG, J., *Data Analysis Using the Method of Least Squares: Extracting the Most Information from Experiments*. Berlin et al.: Springer-Verlag, 2006.

[201] WU, R., SANDSTRÖM, R., and SEITISLEAM, F., "Influence of extra coarse grains on the creep properties of 9 percent CrMoV (P91) steel weldment," *Trans. ASME. J. Eng. Mater. & Tech.*, vol. 26, pp. 87 – 94, 2004.

[202] YAMAGUCHI, K. and NISHIJIMA, S., "Prediction and evaluation of long-term creep-fatigue life," *Fatigue Fract. Eng. Mater. Struct.*, vol. 9, no. 2, pp. 95 – 107, 1986.

[203] YAVARI, P. and LANGDON, T. G., "An examination of the breakdown in creep by viscous glide in solid solution alloys at high stress levels," *Acta Metall.*, vol. 30, no. 12, pp. 2181 – 2196, 1982.

i want morebooks!

Buy your books fast and straightforward online - at one of the world's fastest growing online book stores! Environmentally sound due to Print-on-Demand technologies.

Buy your books online at
www.get-morebooks.com

Kaufen Sie Ihre Bücher schnell und unkompliziert online – auf einer der am schnellsten wachsenden Buchhandelsplattformen weltweit! Dank Print-On-Demand umwelt- und ressourcenschonend produziert.

Bücher schneller online kaufen
www.morebooks.de

OmniScriptum Marketing DEU GmbH
Heinrich-Böcking-Str. 6-8
D - 66121 Saarbrücken
Telefax: +49 681 93 81 567-9

info@omniscriptum.de
www.omniscriptum.de

Printed by Books on Demand GmbH, Norderstedt / Germany